Frank Alper

Our Existence is Mind

Healing Methods for the 3rd Millennium

Copyright © 2016 Katharina Alper
ISBN 978-3-9524451-2-9

Edition ADAMIS

Katharina Alper
Zwydenweg 14
CH 6052 Hergiswil NW Switzerland
Phone: 0041 41 630 33 01

Email: katharina.alper@adamis.ch
Website: www.adamis.ch

Ordering Information:
Special discounts are available on quantity purchases by corporations, associations, educators, and others. For details, contact the publisher at the above listed address.

Editor: Claudia Dale, www.yoursoulscontent.com
Layout: Katharina Alper
Cover design: Katharina Alper
Diagrams: Katharina Alper

A Catalogue recorded for this book is available from the Swiss National Library. Detailed data are available http://www.nb.admin.ch/.

Printed by: Lulu

No part of this book may be reproduced by any mechanical, photographic, or electronic process, or in the form of audio recording, nor may it be stored in a retrieval system, transmitted, or otherwise be copied for public or private use - other than "faith use" without permission of Katharina Alper

www.adamis.ch

Dedication

This book is dedicated to the people all over this World.
I am thankful that they have surrendered their body, mind and soul for spiritual healing. Only through those experiences was I able to write this book. Only through them I was able to develop the processes and techniques which will be written in the following pages.

I feel eternal gratitude for the "universal collective consciousness" – the source of all knowledge. Through those levels of consciousness I was able to gain knowledge, wisdom and create new tools and applications for healing.

Finally I honor my soul, Adamis, for its presence in my life. It has taught me, in these many years of life, the conscious union and the true meaning of grace-love.

CONTENTS

Foreword	XI
Frank Alper	XIII
Introduction	1
Soul and Mind	3
The Soul	5
The Personality	7
The Sub-Conscious Mind	10
Making People Wrong	15
Mind Expression	18
Channeling by Carl Jung	20
Dealing with Terminal Conditions	26
Arthritis	27
Malignant Tumors	32
Case Histories	35
Benign Cysts and Tumors	40
A.I.D.S.	42
Congenital Disease	43
Spiritual Causes of Disease	46
Acne	47
Adrenal Glands	47
A.I.D.S.	48
Allergies	48
Alzheimer's Disease	48
Amnesia	48
Anemia	49
Anorexia	49
Appendicitis	49
Arthritis	49
Asthma	50
Bladder	50
Bowel problems	50
Breast Cancer	51
Bronchitis	51
Bursitis	51
Cancer	52
Cerebral Palsy	52
Colitis	52
Diabetes	53
Emphysema	53

Epilepsy 53
Glaucoma, Cataracts, Corneal Conditions 53
Gall Bladder 54
Genital Organs... 54
Hearing Problems... 54
Heart Disease.. 55
Hepatitis, Liver 55
Herpes.. 56
Hodgkin's Disease 56
Hyper Activity.. 56
Hypoglycemia 57
Kidneys... 57
Leukemia 57
Lymphatic Disorders.. 58
Mental Depression 58
Migraines.. 58
Multiple Sclerosis 59
Muscular Dystrophy.. 59
Myopia 59
Neuromuscular Problems 60
Obesity... 60
Osteomyelitis... 60
Pancreas... 61
Parkinson's Disease.. 61
Pituitary Conditions 61
Polio... 62
Prostate Conditions 62
Psoriasis 62
Rheumatism.. 63
Sciatica... 63
Sinus Conditions 63
Skin Conditions 64
Spine.. 64
Spinal Meningitis.. 64
Spleen 64
Thyroid... 65
Tuberculosis.. 65
Ulcers... 65
Venereal Disease 65
Grace 66

Universal Healing Laws	68
Absentee Healing	78
Healing Energy Aberrations	83
Thought Form Energy Invasion	83
Self–Obsession	86
Soul Invasion or Possession	89
Clearing Process	93
Accidents	97
Karmic Disabilities	101
Physical and Spiritual Healing	105
Spiritual Healing	110
Counseling Techniques	112
Holistic Healing Procedure	124
The Chakras	127
Energy Insertion Points	138
Life Is	147
The Aura	149
Cellular Memory Healing	152
Success and Failure Mechanisms	159
Building Confidence	162
Happiness	174
Unhappiness	174
The 36 Atlantean Chants	176
Gratitude and Dedication	181
Frank Alper: Exploring Atlantis	182
Soul Plan by Blue Marsden	183

Foreword

The early days of my life composed a symphony which, for many people, created disharmony.

As a child I felt different from all the others without knowing why. I was shy, introverted and alone. I was not able to relate to anyone, to feel or to express.
I grew up in my own world and three years of military service, graduation from university, and fifteen years as a business man could not bring any changes.

At the age of thirty-nine I had the feeling that I was close to physical death. I was not ill; I never had been seriously ill in my entire life. But still this feeling became very dominant within me.

At that time I did not know anything about the "spiritual cycles" in life. I had never even heard about meditation. All I knew was that I had to change my life - otherwise I would die.

I allowed this life-cycle to come to an end. I moved and gave myself the space and opportunity to experience a new "birth", and start a new chapter in my book of life.

The story of my spiritual development started to unfold and has lasted all these years. Due to eternal changes and evolution it will never come to an end. I started to research new concepts with great intensity and passion. After forty years I had finally discovered something which was "real" for me. Slowly I began to understand life and eternal existence.

Today, I am in a fuller expression of my life with over 30 years as a teacher, a conscious channel and, most of all, a spiritual healer. I love to see the light shining in people's eyes when they discover the path of self-love. I love to show people the path which leads them to discover their own souls.

I have written this book in a way that everyone can understand it. You will find all of the explanations and descriptions logical and realistic.
We are all the same. We all have the same abilities and the same source ... the Universal soul or the creator.

Maybe I have discovered something which you might not be aware of:
OUR EXISTENCE IS MIND.

Frank Alper

1930-2007

Frank Alper was an exceptional spiritual healer, teacher and channel.
He began his spiritual work in the 70's in Phoenix Arizona.
For many years he held evening classes called "The Carousel".

He introduced the ancient Atlantean methods and techniques of crystals energies and healing.

In the late 80's he expanded his work to Europe which was mainly Germany, Austria and Switzerland and also in the 90's he spend some years in Japan.

As a Universal channel Frank brought many beings, Masters and energies into the consciousness of mankind. One of his first teachers was Solomon. Solomon guided Frank through the completion of his initiations.

One day in the 70's, Frank located the Star Ship "Jupiter 1" with Commander Murvin.
To many of Frank's students, Murvin has become the main Master for structural corrections in the body, such as teeth and bones. The experiences for healing results are usually complete within approximately seven days. Connect with Murvin and you find him to be a reliable brother with a good sense of space humor.

In 1980 Frank introduced Kryon in an evening channeled lecture, called "Moses and the Bible". Some years later, Lee Carroll was "called" to a channeling with Frank, where Lee had his first encounter with Kryon. On behalf of Frank, I thank Kryon and Lee Carroll for opening the spiritual mind to a very large number of people.
Today Kryon is known worldwide and, as time has passed, Kryon has connected to other channels from different continents.

Within the new Millennium the master Aletia, who originates from the same Universe as Kryon, came into our consciousness. Aletia comes to us with female energies. She brought the unusual energy combination of love and magnetic energies. With this very unusual energy combination of Universal Love and the speed of magnetic energies, the process of healing and recovery is immensely shortened. Aletia is assisting within many healing methods, such as "the Aletia - Love - Energy - Transfer" which is applied by many therapists all over the world. She is a valuable assistant within the genetic-energy-reconstruction, which is named ENERGENETICS® and was introduced by Frank in 2002.

Moses not only brought spiritual philosophy but also the spiritual Numerology. This Numerology is quite unusual but very accurate and now is being used and taught by many of Frank's students.
The list of "Beings" Frank channeled is endless and not all are yet widely known.

Each Master has their own specialty and qualities, although any master can be called for anything, one still prefers to call a specialist in specific areas.

One of Frank's most important goals was to help each individual become free of emotional boundaries and achieve the state of freedom of soul existence, and to integrate both aspects in balance for a more fulfilled life. During more than 30 years of spiritual work his connection to the Universal existence greatly expanded. Countless people all over the world studied his spiritual philosophy, knowledge and great techniques and now carry on with his wonderful work.

Frank always made the statement that we, as human beings, are not here to have spiritual experiences, but we are spiritual beings who are here to have human experiences.

In 2005, Frank began his "final chapter" as he called it. His faith in God and spirit was solid and unshakable; but with this faith he neglected the creeping uneasiness in his physical structure. Deep within he knew his days had been counted and would come to completion. Within the balance of his years he started to teach in a restless hurry and conducted his last seminar in October 2007 in Switzerland.

His passing on December 7, 2007 was a shock and confused many, but Frank always said: "You never ask spirit "why" and you just accept the will of God".

Introduction

The ancient practice of energy healing through "hands" started the moment when Adam and Eve touched each other and thoughts, feelings and emotions were discovered.

Over time, the formal religions developed and the healing with "Gods power" became an integral part of the religious act.

Today, the metaphysical beliefs have started to develop into esoteric and spiritual philosophies. The art of energy healing has brought the understanding of soul and made it acceptable. We discover that we are single-standing-beings, who are influenced deeply by the conscious and unconscious actions of the body.

How long will we point with the finger of responsibility to someone else? We have to look at ourselves in the mirror, open our mouth, and point the finger towards ourselves: "Only I am responsible for me and my life."

If our existence is "mind", then what is "mind"? Is it our physical brain, our thought patterns and the experiences we have learned? Is it the reactive processes of our personality; or everything together? Maybe our "Mind" also holds unconscious collective energies, which apparently do not have a connection to our conscious behavioral patterns.

A definite interpretation of mind can only have a statement of many possibilities. For our purpose here I call "mind" the energy of the entire collective existence, which is within each one and is consciously available. Therefore, this boundless energy knowledge and the power of expression can be developed. If we can learn to intensify this energy, if we can create and guide the power of expression, then we can also strongly influence our three aspects in life - body, mind and soul.

The word "healing" means to supply the body with energy so it can cure the disturbance. Each one has to take the responsibility and accept that the actual healing process is something between himself and God.

Throughout the entire book there are examples for "healing" processes and also long-term illnesses. The purpose of this description is to become aware of the possible spiritual causes of disease and to bring awareness to the influences in life.

This book and the information have no scientific bases. It is a philosophy of life and being. There can never be a guarantee that somebody can be cured of his illness.

This book does not contain any continuous literature or bibliographies. I became aware of all the information through my mind. There are no final statements. Everything is influenced and expressed through variable factors of choice which one makes during the course of one's life.

Welcome to your mind and to the world of healing. Join in on the exploration of the conscious and unconscious parts of life and start the process of integrating them into the full expression.

I lead my life in the service of all who search for their truth. It is my goal to support people in finding the eternal beauty of their soul and to lead a life in health and joy.

Soul and Mind

Almost every written language tells us that we are created in the image of God. What image? Man, woman, white, black, yellow, or red?

Perhaps we might consider that God is a soul of the greatest magnitude conceivable, whose energy essence permeates the expanse of the universe. Everyone must decide for themselves and live by their decision. Whatever your concepts are, they always serve you as you progress into the final truth of your soul.

When I was a small child, an uncle of mine passed away. After the funeral I asked my parents, "What happens to him now?" They told me that his soul went to Heaven to be with God. I accepted that response but inside I did not understand. I had no conception of what they were talking about.

How about you? Do you ever think of your soul or are you just too busy with life? Is a soul just something that we have labeled, categorized, and put on a shelf so we do not have to explore the concept?

Let's open our minds and explore. The Universal Law states: As above, so below. As below, so above. The interpretation of this means that existence or history always repeats itself. Whatever occurs in one dimension always occurs exactly the same in all dimensions in ascending or descending order.

In our existence on Earth, a single sperm fertilizes a single egg creating life. The union begins to nourish itself from its universe, the mother. Division and multiplication begin and, at the proper time of maturity, the mother universe gives birth to a child.
We travel from below to above. It is my belief that when God creates a soul, a union between the essence of God and a single energy cell of the Universal Mass called the Central Sun takes place.

This cell union begins the process of nurturing itself from the Mother/Father God in the same manner as the fetus in the womb. When this activated cellular structure reaches the required energy frequency of existence, a soul has come into being.

Are our souls, human souls, any different from all other expressions and levels of souls on Earth? I believe that we are different. I have reached this conclusion as the result of energy involvement with the soul expression of many people during the last twenty-five years.

Through my experience, I have come to the understanding that the human soul contains the energies of existence that enable it to express conscious, verbal intelligence. It also has the capacity for logical reasoning by utilizing mind decisions based on past experience. This ability extends back to prenatal life as well as to the reactions that occurred during fetal development.

We have three levels of consciousness inside our physical bodies. They are all distinct, separate levels or aspects of consciousness expression.

1. The soul or universal truth expression. This is often referred to as the higher consciousness.
2. The personality. This is composed of the ego and conscious expression of experience, conditioning and knowledge.
3. The sub-consciousness, or reactive, memory storage, and response center of the mind and body.

The Soul

The soul or higher consciousness expresses itself through energy impulses directed to the part of the brain associated with conscious thought or impressions. It is through this process that the soul is able to subtly allow the personality's mind know what actions and expressions are in accordance with the soul's purpose and goals for the current lifetime.

This process is often implemented by mind impressions of the following:

a. Thoughts
b. Sensory feelings
c. Dreams
d. Spontaneous ideas
e. Visions
f. Fantasies

These processes constantly flow in and out of the mind. We disregard most of them and seldom pay attention to our own thoughts and feelings. We seek the advice and counsel of others rather than take chances on the unknown and untried actions.

Your soul is here to evolve and to serve others. These processes are the only mediums through which your soul can communicate with you and make you aware of its innate plan for your lifetime.

Let us consider something for a moment. Human beings are created in perfection. Look at the results of the union between one sperm and one egg. God is truly miraculous! This can raise several questions. In this marvelous creation of billions of cells, why would God create a brain that remains 80% dormant? Why do we utilize only one third of the rods and cones in our eyes for our visual perception?

Case Example 1

Many years ago, a twenty-eight-year old man came to see me. He was 98% blind from birth and yet operated a successful newspaper stand. He wanted to learn about energy and healing.

As the class progressed, I taught him how to begin to activate dormant energy centers in his brain. These centers began to stimulate the associated dormant rods and cones of his eyes.

In one month this man could see. Not as you and I see but he was able to perceive color, energy patterns and visions with his eyes closed and blind. He developed the ability to mentally project energy to someone and see the organs in their body. From this, he was able to relate conditions of health to them verbally.

In time, he developed the ability to identify energy vibrations of people he knew by sensing the energy differences in different people.

Case Example 2

In 1978 I was in San Francisco. A man brought his wife to see me. She was in a wheelchair, the victim of an automobile accident. Spinal damage had paralyzed her legs and body. The woman had been an energy healer for many years. As a result of the accident, she developed severe energy blockages in her spine and was unable to control and disseminate energy to others for healing.

I told her that I was going to try to create a new energy meridian outside of her spine that connected to energy insertion points on her shoulders. If it were successful, she would be able to resume her healing practice.

The process of brain activation and mental control took three months before any measurable results appeared. Within six months, her intense desire and dedication had produced the desired results. She is now totally in control of her energy system and has resumed her healing practice.

How did these so-called miracles happen? I do not consider them to be miracles. To me they were the result of the systematic activation of previously dormant brain cell tissue. OUR EXISTENCE IS MIND.

The Personality

The personality or ego expression is the active, expressive mechanism of thought, decision, and action that creates truth or lies and sets the stage for reactions.

Most of the time, we ignore the subtle input of our soul. We base the greater portion of our conscious actions purely on our emotional reactions or responses. Rarely do we pause and try to feel if a contemplated action is our truth and valid for ourselves. We consistently plunge into the peaks and valleys of emotional energy. We often enjoy riding the roller coaster of indecision and insecurity.

How often do you ask yourself:

- Why didn't I listen to my good feeling?
- Why didn't I do what I wanted to do instead of asking someone for advice?
- When am I going to learn that I know what is best for me?
- Why don't I trust my own thoughts?

How many times you said:

- I went to bed with this unresolved problem and the moment I awoke, the answer was in my mind.
- During the night while I was asleep, I had this great idea come to me in a dream.
- Look at this, isn't it great? I don't know where it came from. It just popped into my mind.

Don't ask me why? I just know:

- This is what I have to do.
- This is where I have to move.
- I have to call this person.
- I need to read that book.
- I just feel that I know you.

Each of these thoughts or statements should make you aware of the presence of your higher consciousness or soul. If you begin to gradually acknowledge these subtleties, the connection and awareness will increase in power and the frequency of occurrence.

One of the prime functions of the ego mind is to distinguish and act from three areas of emotional stimuli. These are: Desire, want, and need.

Desire

A desire can be defined as a pure emotional response to a previously not experienced person, situation, expression, or thing.

Want

A want is created from the repeated exposure to a desire that have elicited a positive, emotional reaction from the person.

Need

A need is determined by examining the want and validating it as a necessary acquisition for the expression of your life.

Example

One morning you walk out of your home and see a shiny, new car parked next door. Your response could be, "What a great car, I wish I had something like that".
You have just expressed a desire. You had no past experience with that car, just the initial, visual perception.

Every day, when you leave your home you see that car. It finally affects you as the result of a buildup of positive emotional responses. You say, "I want that car."

That is a valid response to that situation. You have constantly repeated the exposure to the desire and it has not left you. In fact, you have created a positive, conditioned response to the car.

What happens if you overreact and say, "I need that car?" In this case, you would have stated an invalid need, and could create serious complications for yourself.

What if the price of the car is beyond your income? This would generate a situation of failure and unworthiness.

What is the valid need? The valid need is transportation. Anything extra must remain in the category of a want or desire. In that way, if the item is too expensive and beyond your means, you will not become involved in the reactive energies of failure and unworthiness.

If we fail to obtain our wants, we may experience sadness or disappointment, but not failure. And it will not become recorded in the subconsciousness. It will remain a pure conscious, emotional reaction.
This is very important to understand. Every time we fail to fill a valid need, we become failure oriented. This is out of order when the failure is not valid.

If we will remember that everything is always in a constant state of change, we will not develop expectations. We will remain open to the development of new needs, new wants, and new desires. This will enable us to avoid situations of sacrifice and express the full flexibility of life.

The Sub-Conscious Mind

The subconscious mind or reactive system is the computer storage center for your body and mind. Here are recorded all the reaction impulses that are generated from every thought, statement, and action initiated by the personality mind.

- Likes and dislikes
- Joys and sorrows
- Worth and unworthiness
- Judgment and acceptance
- Love and hate
- Success and failure
- Happiness and sadness
- Grace and expectations

These are some emotional responses for actions that you initiate, whether you complete them or not. The sub-conscious mind emits the energy reactive response, which becomes transmitted to the body cellular structure and indicates the truth reaction to that specific action.

For example: If you are bitten by a snake and respond in fear, the recorded truth becomes: "I am afraid of snakes. Snakes bite."

Our main areas of focus are the physical, emotional and mental effects that are created from the result of conscious statements that become recorded in the sub-conscious computer as a negative or unworthy expression.

The sub-conscious mind does not distinguish between what we call truth or lies. Its role is just to record your statements of thoughts, feelings and reactions. This creates our pre-conditioned responses to people, things and situations based on our past experience of similar circumstances.

It is through the use of the sub-conscious mind that we can begin to effect changes in the old patterns of conditioned responses that no longer contribute to the fulfillment of our lives.

OUR EXISTENCE IS MIND. The patterns of energy we can create from our mind will become the truth of the sub-conscious, and the new physical, mental, or emotional expressions.

Example:

The mother, whose child has not lived up to her expectations and takes drugs, alcohol, etc. will often blame herself for the condition. She might tell herself, "It must be my fault. I wasn't a good mother." If she keeps reinforcing this statement of mind, it will become her sub-conscious truth. If this occurs, her body may begin to create the actual, physical conditions to make the mind statement the truth. Her body could begin to develop alterations of conditions in the related areas of her body. Most often, this will occur in the breast or genital areas — the distinctive parts of the body that relate to motherhood and womanhood.

If the mind states that one is unworthy and is being judged, the body may begin the destruction. Breast or vaginal cancer could manifest in either of these areas.

Case Example:

A mature mother with three children asked me why she was consistently developing small tumors in the palm of her left hand. At that time, she already had seven surgical scars in that area.

In counseling, we went back to her childhood to the time when she was six years old. She began to cry hysterically and called her left hand, "a bad hand that was punished by mommy." We brought her back to the present, calmed her down and asked her to relate the details of the incident that she had blocked out of her consciousness for thirty years.

This is what she recalled. "When I was six years old, my mother saw me touching my private parts. She screamed at me. She called me many bad names, took my left hand and held it over the gas stove to burn away all the bad that was inside of me.

Now we began to understand the condition. For thirty years the sub-conscious had been programmed: our left hand is very bad and unworthy. This reinforced statement from the mind created distorted energy patterns that led to the creation of tumors in the left hand. The hand was systematically being destroyed, operation after operation. Scar after scar.

Several months later, the hand had begun to heal itself. The program was negated. The mind input to the sub-consciousness had been eliminated and replaced by positive affirmations.

She began to love her left hand. She cared for it. She acknowledged its presence and value for her in her life and, for the first time, began to use the hand for constructive purposes.

Other areas and sections of our body express many conditions created by reactions from the consciousness. For example: An expression of disease in the left leg indicates an insecurity of expression. The repetition of past actions creates safe spaces for us. They have a knowing of the results of the repeated action. Many people want to remain anchored in place. This is how they can remain safe and secure without any risk in their lives.

The throat is called the vocal, expressive center of all aspects of our personal expression.

For Example:
- Words become stuck in the throat.
- Rather than say something to hurt someone, I say nothing.
- So what if I don't tell them how I truly feel. I just want to keep peace.
- If I comment about that, they will just laugh at me.
- Sometimes I tell lies so they will like and approve of me.
- I would rather suffer in silence than face the consequences of an unknown reaction.

Every person keeps thoughts locked in their mind. They are safe there. As long as we do not express them, we can make believe that they do not exist. Everyone says, "Maybe they will just go away after a while." What does this repression create inside of us? Does it really help to remain silent?

When the expression of truth is suppressed, a block can appear in the energy circuitry of the body. This could cause one or more of the following conditions.

- Feeling like you have a lump in your throat, causing you to constantly clear your throat.
- Developing throat polyps.
- Being susceptible to sore throats or hoarseness all the time.
- Throat cancer brought on by the silent statement:

"I do not speak my truth. I speak what others want to hear and sacrifice my truth for them. My throat is unworthy to express my truth."

Our existence is the programming of our mind!

From the moment a child is conceived, the patterns of sub-conscious reaction begin to formulate and become stored in the unconscious computer.

Maybe you might ask yourself:

- Was I conceived on purpose or by mistake?
- Was I conceived in love or anger?
- Was I conceived from desire, want, or need?
- Was I isolated after birth or kept with my mother?
- Was I premature or longer than full term?
- Was I born breach, reverse, or with forceps?

There are many questions that could be asked, and each of them relates to the actual circumstances, and the reactions of your sub-conscious mind. With each reaction, an energy pattern is created in your reactive mechanisms that subtly direct your conscious behavior.

As a child, were you told that:

- You will never amount to anything.
- Children should just listen, not speak.
- Boys never hug or kiss men. Just shake hands.
- Girls should be suspicious of all men.
- Don't ever touch yourself. God will punish you.

- Sex is dirty and evil.
- You are ugly. No one will ever want you.
- Why aren't you like your sister/brother?
- I wanted a boy, not a girl or girl, not a boy.

As a child, did you ever experience?

- An insufficient amount of love.
- Rejection at your attempts to be affectionate.
- Abandonment.
- Physical, metal, or emotional abuse.
- Lack of recognition of your achievements.

All of the statements we have listed contribute to the programming of the sub-conscious, automatic response systems. No one has escaped. Everyone can relate to at least several of the aforementioned statements. We have all been a victim of some form of dysfunctional behavior patterns either before or after birth.

Most of the time, we are not even aware of these conditions or their causes. We have blocked them out of our minds a long time ago. What remains are the automatic, unconscious reactions that reinforce this type of programming, and continue to hold us back from living full and joyous lives.

Remember, every statement of your conscious mind creates a sub-conscious condition of response to make the mind-thought truth, never a liar. With these life conditions, we need a method to assist us in replacing these deep-seated patterns with the new truth we want and need in our lives.

Making People Wrong

To a young child, mommy and daddy are as God. They provide for every need of the child and are the first people to express love to the child. Because of these circumstances, whatever action occurs that is not in order; the child will always assume that it is my fault. Even if the child was sexually abused, they will blame themselves. After all, mommy and daddy are as God.

One of the most powerful counseling tools is to have the person make the statement, "Mommy or daddy was wrong. They made a mistake." These words begin to free the individual from self-judgment and punishment. As long as they blame themselves, healing becomes a most difficult process.

So, if you can allow people to be human and make mistakes, then you can:
- Take back all of your own power. You don't have to subconsciously take improper actions just to make those people right.
- Say to them, "I love you, but I am not you. I am me."
- Say that your truth does not have to be the same as their truth.
- Say that just because you disagree does not mean that you do not love them.
- Acknowledge that you were abused.
- Admit to yourself that you were not wanted, were rejected, and never understood or were allowed to express yourself.

In order to accomplish change, you must first admit to yourself and accept that whatever happened to you really happened. It was not an illusion. Without the admission, you are still lying to yourself, and any change will only be verbal and not take effect in your reactive systems. If you can make the admission, you must then state: "I accept that all that has happened to me was real. I no longer have a need for these matters in my life."

This statement of allowance automatically allows people to be wrong. You do not have to confront anyone.

You do not have to yell or scream or involve yourself in an emotional outburst of anger. The statement will make you free, and allow you to take back your personal power and to be in total control of your life.

The next action is to begin to re-program the sub-conscious computer. We do this through replacement techniques, not by methods of release. Releasing means: letting something go out of you. If you could accomplish this, you would create a void inside, an empty space that could create insecurities in you. Something will fill the empty space and you do not know what will come inside. The process of releasing can cause resistance and make change difficult to achieve.

When we speak of releasing something, we take the attitude that it will leave our body. This is a misconception. The true process of release involves taking an active pattern and making it passive. By this process, the pattern only serves as memory to help you in future decisions and wisdom.

Energy is eternal. That means that any pattern remains with you for life. The key is choice. Which expression do you want to be an active part of your life?

Replacement techniques involve the conscious re-programming of your computer mind. When you have compiled a list of all the wrongs imposed on you, make a list of how you want your life to be. Begin to take actions as if you are already that new person.

Establish short-term goals for yourself. Make them easy to reach and begin to experience your newfound successes. In a short time, you will have begun to push out all the old patterns and a new truth will begin to replace the old pattern in your sub-consciousness. Change will come without any fear. You will be able to look back, smile, and wonder what happened to the old you.

One problem might arise. No one wants to make their parents wrong. We have not been raised that way. We know that if we want them to love us, we must never argue; and do what we can to make their words our law.

At the present time, you have reached the age and level of intelligence where you can finally be responsible for your actions all by yourself. You now know that everyone makes mistakes some of the time, even your parents. No one is ever perfect. No one!

Mind Expression

The great majority of people lead their daily lives maintaining a division between their minds and their bodies. How often do you try to feel what is going on inside your body?

In order to achieve unity, instead of separation, try to periodically follow these guidelines:

1. Before you plan your meal, take a deep breath, silently move your mind down into your body and sense what it needs at that time.
2. Before you make a decision, move your mind down to your solar plexus and feel if the decision is compatible inside, or does it create uneasiness.
3. Before saying, "Yes" to any proposal, close your eyes and ask yourself, "Is taking this action in alignment with the truth of my soul, or is it an emotional desire?"

Once a week lie down for ten minutes. Breathe deeply through your mouth and consciously move your mind down into your physical body. Begin to explore all the emotional reactive centers and feel what has become stuck there. These areas are the genitals, stomach, breasts, heart, knees, and upper thighs. By doing this, you will locate anger, fear, failure, and resentment energies that may have accumulated during the course of the week. Once you have identified them, you can replace them from your mind and become free.

Our body is the expressive tool of our soul. Our mind must be aware of the conditions that exist in our body to maintain health and joy in our lives. If it does not, we can become dysfunctional and susceptible to disease. Every organ, all the so-called vanity areas of your expression, and every cell of the body compose your attitude towards your body.
Affirm: My body is the temple of my soul. I honor every organ in my body as a worthy creation of God.

Do you ever find yourself saying?
- I hate my body.
- I wish my breasts were larger/smaller.
- I would like to cut twenty pounds away from my thighs.
- My nose is too big/small.
- I'm really ugly.
- I wish I had hair on my chest.
- I hate my freckles. I got them from my parents.
- I hate sexual activity.

As long as we keep our mind attached to our conscious thought patterns, we can continue to avoid the real truth. We can rationalize, justify and make believe. The moment we move the mind awareness down into the physical body, we are capable of freeing the mind from thought. We can then allow it to sense and feel the energy reactions in the chakras and organs of our bodies.

Our truth is inside the organs and chakras. All the reactive patterns deposit themselves there. They begin to be the seeds of dysfunctional, responsive behavior. In order for anyone to alter a response pattern, they must first acknowledge its presence. This does not mean that you have to suffer and endure any emotional pain or distress. The key is to identify and acknowledge the presence of the hurt or pain. Once this has been done, you can begin using the replacement techniques we have described in the healing chapters to permanently re-program your responsive systems.

Medical science has acknowledged that the mind has the capability to continue to expand in power, concentration, and focus regardless of age. There is no credibility to statements of a weak mind because of the age factor. You are the expression of your mind!

Channeling by Carl Jung

Part 1

We are Evoran, the soul of one who walked upon your planet as Carl Jung. And we have requested to speak to you to perhaps bring a concept into your consciousness that will begin to free you from the psychological destructive patterns of society. When we physically walked upon your planet, our awareness of consciousness in relation to our soul was in totality. The times we attempted to bring this truth to mankind were continually rejected by our peers and by most of the medical community. And so, we have made the decision that in the future our service will be better for people from our true expression as spirit. And unless a strange circumstance appears, we have no plans to walk upon your planet again. The traditional psychiatric and psychological therapy is slowly beginning to realize that permanent effectiveness is most difficult for them to achieve. Those who remain in their expression of integrity have begun to search to explore what you refer to as alternative methods dealing with the cellular structure and the soul patterns of energy response.

As you look around yourselves at those of you that serve in the form of counselors, you have begun to notice the difficulty that exists in the opportunity for two people to create and maintain a nourishing relationship, and this is our topic of discussion for today.

As one continues their spiritual enfoldment, patterns of energy come to consciousness, patterns from your soul. The truth and knowledge of your soul. And all of a sudden your emotional expression does not seem to be as truthful as it has been in the past. It seems something has changed. And this is the key for the future.

Of the many karmic experience patterns of life, the one that is most difficult to understand is the emotional pattern of expression. For here we have the causative factors of the turmoil inside of all of the reactive patterns that cause disease and the psychological insecurities of people.
And so today, we shall pose words to you for you to discuss among yourself, for you to discuss within yourself.

I cannot say to you when it is the right time but certainly it shall come to be that the emotional patterns of relationships truly shall change on your planet. And the emotional reaction and response shall take, shall we say, second place to the purity of soul energy response. The formalized situation of two people together shall quite slowly disappear for it becomes a limiting vibration of relationship. It has nothing to do with a condition of moral or social structure. It has to do with the truth of universal soul and consciousness. The sensitivity of energies that you are going to utilize in your new millennium requires a finite compatibility with your emotional patterns. When one enters a relationship of emotions, the great majority of the time one looks to the other to be the vehicle that will fill their own voids. It is like you say to yourself, "How wonderful! They will replace what is missing in me." And from the moment of the beginning of that attitude, the relationship is doomed to failure for it signifies that each one is going to be a victim of the other.

Where do you nourish yourself? Where is the place that can fill your voids? Only from the frequencies of inner soul. And, therefore, it becomes of the greatest necessity that each of you establishes a relationship of love with the frequencies of your soul. Here is your power. Here is the clarity that you can utilize in the continuing path of life. You will not find this or receive this from other humans. And once this experience of love has been achieved within your emotional body, all shall change. You shall enter the position of detachment. And you shall know that the pure love only comes from soul; you cannot experience it from humans.

Within the human emotional pattern is the variable to reactive energies that creates the karmic expressions of exchange. That is the pattern of life upon your planet.

Your soul exists within the pattern of the universe. And you have the capacity to become part of this expression. Perhaps it is difficult for you to understand, but the time shall truly come for you and for the spiritual awakened child that you shall no longer share your physical bodies with one purely for the exchange of emotions; you shall learn to honor and utilize the sharing of structure as to complete the sharing of soul. And the angels shall sing for you. And your growth shall increase.

And your physical structure shall heal. The psychological damage that your society has placed upon you shall be replaced by the understanding of life.

One might ask, what is life? Is it an existence upon this planet? I do not have an answer for you. It depends upon the place you have your consciousness. It depends upon the place you wish to put your emotional expressions. It depends upon your values within your society. It depends upon your relationship to your soul. How many realities are there for you? And what is reality? Is reality something that you can count on that is an anchor for you; that shall always be there? And, if that is true, then your realities are soul and God. And the rest are frequencies of constant vacillation to change all the time. They change as the thought patterns you produce change. They change as you are affected by the collective consciousness. They change as you are affected by the actions of others.

Life is soul. Soul is God. Life is the universe. And within the purity of the highest existence the only vibration of reality is love. And, if you cannot achieve one other expression within the total span of your physical incarnation, experience the truth of love. And all other things in your life fall into the proper placement of importance for you. And all of a sudden, it does not matter what kind of pretty clothes you are wearing. It does not matter the size and projection of your physical structure. It does not matter the amount of finances that you have. All that matters is the love. The healer of all things! My children, life is eternal. There is no death to fear. And yet upon Earth, there is death. And people rush to finish things while they are there. You do not wish to miss out on anything. But anything here is nothing. A fleeting moment of expression that the moment that your soul is in its purity, is nothing at all! And so, you shall re-examine all of your values. The changes shall come in a gradual manner of expression, but you must begin to look at your life from different dimensions. And when a discovery is made for you, try to utilize the teachings that Adamis has expressed to you, never to create psychological voids inside, the greatest mistake of mainstream therapy.
Create your new experience and replace the old with the new and then you shall walk in truth, the love of your soul shall become one of peace and joy for you. Within the cells of different areas of your structure are contained vibrations of past experiences of your soul as well as your collectives. You call this karma.

And for quite a few years of your lives you are consciously affected in your actions by the presence of these energy patterns. These are in addition to the exposure, as a child, to the action energy of your family structure. Today it is time to be free. Today is your tomorrow. Let it go.

Create the new experience for yourself. Embrace existence of soul. Acknowledge your universal existence and the power shall come to your mind. And your truth shall become evident to you.

These are the words we have selected to share with you at this time. We are pleased to connect with you. And we wish you continuing joy and truth. We take your leave.

Part 2

Our greetings to you: When the essence of soul elects to incarnate in a physical life upon your planet it involves itself within the selection of the mother and father. The major reason for this selection is the psychological and environmental conditions that the child shall be confronted with during the formative years of its personality. The great majority of time this choice is made in relation to the obstacles that the child shall be confronted with. One must understand that these obstacles have not been selected as a form of punishment or any manner of destructive nature. All is in a mathematical proportion that directly relates to the energy evolution of the soul. The more the soul has achieved a degree of evolution, the greater the potential of energy service for God. Therefore, the refinement of the psychological processes becomes of greater necessity in relationship to the degree of conscious service, and consequently the obstacles become greater. Remember the laws of balance and counterbalance. The greater the frequency of light, the more intense the frequency of dark, and this transfers itself to the obstacles the child has to face in life. One cannot ignore the psychological aspects of spiritual expression. The two are closely integrated and have a joint effect upon your conscious ability to connect to the spiritual dimensions that are the teaching energies of your universe. Perhaps we better reword this for you. You sit here and you strive and work towards a clear communication with your masters.

In order for this to achieve completion, the channel must be clear, for one cannot expect the master soul of evolution to alter its frequencies, for that certainly would cause distortion in its communication to you. And so, creating the balance between your conscious psychological profile and the spiritual peace of your essence allows and creates the clarity of your channel. I hope that is clearer for you.

The subconscious patterns of your emotional frequencies are the area of your expression that requires the most attention. Within these recesses, it is quite normal for mankind to utilize this energy in a form of emotional self-punishment, or as a justification to remain in the past because that action and knowledge has already been experienced. When this condition occurs, many times your conscious psychological profile creates an alter expression of energy. Sometimes this is called the alter ego, something that does not really exist. Many times it can serve you probably to help you determine the proper future decisions and actions. In addition, if you have been conditioned from your family circle, you can utilize this alter expression to justify avoidance of the future, and to create a comfortable hiding place for yourself, remaining upon the treadmill of nothingness. This creates a great danger for when one finds comfort and peace in this manner of expression, one, all by themselves, can generate a split personality and take their conscious expression and hide, so that they do not have to deal with the responsibilities. This situation is something that you shall encounter with many who come to you for therapy. In your applications of energy, you shall develop the sensitivity to detect and locate many of these hidden pockets of energy that are concentrate within the lower centers, and use your light to stimulate a release by accentuating to the conscious mind the choice that you have made available to the individual: to remain hiding in darkness, or to walk in the truth and expression of soul.

The psychological community is rapidly becoming aware of the ineffectiveness of their treatment to create permanent conditions of change. Their methods of total dependency of their clients quite often results in more destruction than was there at the inception of treatment. Gradually, those that remain in their integrity are seeking information and procedures of alternative energy to assist people in changing their conscious patterns of expression.

And so we encourage you to continue according to your present path of exploration. The comparison of experience of replacement of energy will inspire the client to continue and maintain the new attitude of life. And they shall begin their path of discovery. And all of these patterns of expression utilizing energy shall indeed become the operational treatments of psychological disorders in the future. Well, that is all for today.

Dealing with Terminal Conditions

The mention of the word cancer causes people to cringe with fear and change the topic of conversation. We make believe that such a terrible disease only affects other people and, if we do not discuss it, it will go away.

The time will come in the future when the disease cancer will become a distant memory. When? As soon as people live in a condition of self-Grace to themselves as well as to others. When we no longer have a need for an expression of disease, it disappears from our societies.

For purposes of our discussion, it is necessary for us to incorporate certain expressions of disease into a general classification.

They are as follows:
- Arthritis
- Parkinson's disease
- Hodgkin's disease
- Multiple sclerosis
- Malignant tumors
- Benign tumors
- A.I.D.S.

Each of these expressions of imperfection is holistically classified in the same category of causative behavior. All of them are capable of destroying the body, the mind, and eventually generate damage to soul energies. Each of them expresses a major form of destruction, and the rejection of the total perfection of the physical being as a creation of God.

In our review of these expressions of disease, we must make a strong separation between congenital disease and the acquiring of disease during the course of conscious life. For now, we will confine our comments to consciously acquired expressions of disease.

Arthritis

One of the unusual aspects of arthritis is the fact that it can be detected in the auric, or energy field around the body up to ten years before it physically begins to affect the body structure in an active expression. In the auric field it appears as energy hot spots that indicate blockages of the normal flow of energy through the meridian structure. These blockages normally appear in the following places:

- Shoulders
- Elbows
- Wrists
- Fingers
- Hips
- Knees
- Ankles
- Spine

These are the areas that are most commonly affected by this expression of disease.

What causes arthritis? There are many explanations, and we are not going to invalidate any of them. We are just going to suggest some possible causes based upon our years of experience in dealing with this condition. Arthritis is considered to be potentially an all-over-body condition. Even if it appears in one confined area, it has the potential to affect other areas of the body.

According to spiritual law, arthritis may be caused by the evident lack of one of the ingredients of nourishment necessary for the complete expression of what the individual feels is required for their full life.

During the last decade there have been many very sensitive and evolved souls coming into a conscious life on Earth. When we look into their eyes, they seem to be old with wisdom. These children, because of their evolved sensitivity, require a greater amount of love-security than most other children.

The years between two and five are the most important in relation to exposing a child to love. This includes hugging, affection, holding close to create a sense of great security so the child will set a pattern of being loved into its sub-conscious programming. We must realize that by the time a child is five years old this pattern of love or absence of love has become established for life.

Most parents feel that they are giving their child all the love they are capable of sharing. What would be the effect if their capacity of love falls short of the needs of the child? This becomes a truly sad situation. The child may begin to feel unloved even though the parents truly love the child.

We all remember the times during our young years when we became ill. We received so much attention, presents, extra treats, toys, etc. Those were the times we knew that mommy and daddy really loved us! With this joyous memory inside, if the child has a strong need for more love from its parents, it could unconsciously create a disease to draw attention to itself.

Under normal conditions, the child will never blame its parents for the lack of receiving love. It will always blame itself! The sub-conscious programming will be: I am unworthy of love. I am not good enough for them to love me.

Often a child will become angry and bitter. It will begin to resent adults, as it is the adults who do not express love to them. If the anger and the unworthiness remain for a prolonged period of time, the child could develop rheumatoid arthritis. This expression of disease will serve several functions for the child. It will be noticed, draw sympathy to itself, and express the unconscious confirmation of its unworthiness for love.

The spiritual causes for the expression of arthritis in mature years can be the result of prolonged anger and resentment against others, as well as against themselves. The result of this creates patterns of unworthiness and judgment of self. When this occurs, the body begins to alter the inner energy circuits, making the physical structure susceptible to arthritis.

Why arthritis? The nature and form of this disease is to create blocks in particular joints of the body. These blocks can become calcifications impairing full movement and a full expression of activity.

What better way to express unworthiness than to be unable to walk forward, to reach forward, and to be unable to experience new challenges and joys of life? The more the anger and bitterness increases the more the body becomes rigid and immobile until it is unable to become involved in any form of physical expression.

No one consciously wants to do this to themselves, and yet they do. No one wants to become crippled and immobilized with pain, and yet they are.

OUR EXISTENCE IS MIND! And under all conditions your body will make the statements of your state of mind its truth.

When someone comes to see me with a condition of arthritis, the first thing I ask is, "Who are you angry with and for how long?" During the last twenty years I have never had anyone say, "I am not angry at anyone, not even myself." This brings out an important point.

All energy healing of arthritic conditions must be preceded by extensive counseling. If this is not done, it will be most difficult to effect any positive results. The person MUST be made aware of what we call a possible cause of the disease. This can only be determined through counseling techniques. Once the individual becomes aware of the circumstances that could have created the condition, they are faced with a most important decision. Do they NEED the disease any longer? Are they ready to let go and be worthy?

If they cannot answer these questions, they will not be able to relieve the existing condition, as the body will continue to express the programmed input from the past.

Let us assume that a person has come to a full understanding of a possible cause and wishes to change their attitude, as well as their physical condition. Now we can begin the healing process.

Arthritis causes calcifications in the joints of the body. These create energy blocks and sometimes are responsible for the resulting pain. Before we begin any healing process, we determine what must be done to alleviate the condition. With arthritis, our purpose is to gradually break down the calcification and re-establish a normal energy flow in that area of the body.

We are involved in this situation as a spiritual healer, not as a person who cures disease. Please remember this at all times. Our role in working with arthritis is to restore the normal energy flow of energy inside the body. If we keep this in mind, we can focus on our proper role in our service.

When a block exists, causing a stoppage of energy, we insert energy into the body from both sides of the blocked area. For example: if arthritis is present in an elbow, we insert energy from the shoulder and from the wrist. These areas have energy insertion points and conduct the energy directly into the body meridians.

The healer must establish the thought conditions in their mind that they are sending energy from the shoulder to the wrist, and back again to the shoulder. Mentally state that the purpose is to re-connect the energy circuit in the arm. Picture yourself as a set of battery cables pouring energy into the full length of the arm.

In several minutes, you will begin to feel a pulsation in the fingertips of both of your hands. When this occurs, it indicates that the energy has moved through the block and begun to re-establish a normal flow through the arm. Continue the treatment for at least twenty minutes and repeat several times a week until some improvement has been achieved.

One of the unusual side effects of healing arthritis is that the person may experience twinges of pain for a short period of time. If this happens, tell them that it is caused by the energy breaking through the calcification and is a positive reaction to the healing process.

The same basic principle applies to healing arthritis located anywhere in the body. This is, to create a circuit on both sides of the afflicted area, and the energizing of the circuit with both hands.

When dealing with cases of chronically advanced arthritis, such as deformed fingers, diseased hips or knees, you must be realistic. In such advanced cases you are rarely going to be an instrument in assisting total recovery. If someone comes with advanced arthritis, the most you can hope to accomplish is to alleviate pain and help them from an emotional standpoint.

Through counseling, you may be able to assist them in halting the further progression of the disease. If you can serve them in this capacity, you will have done them a great service in their life.

Example

A man brought his wife for healing of an extremely advanced case of arthritis. She was so badly crippled that she needed assistance in walking. The man had retired early from his job to stay home and care for her. Her speech was so badly impaired by the arthritis in her jaw that he had to speak for her.

The woman was in her early forties and had been severely impaired for fifteen years. I began to ask her husband an extensive series of questions regarding her life and experiences during the past twenty years.

We were dealing with a very gentle woman. He told me that she had never complained during the course of her life. She had been always ready to assist anyone who needed her, putting aside her own needs to help them. It came out that during the twenty years of their marriage, she had never argued, disagreed, or become angry at anything that had taken place.

I began to speak to her and told her that it was finally time for her to feel and speak about what she had been keeping inside of herself for so many years. I told her that she did not need to be perfect. She had a right to be imperfect and human like everyone else. She had led a life of total sacrifice to others, subjugating her own truth, her needs and desires, to please others and not make waves.

At that time her condition was too advanced for any type of healing that might lead to even a partial curing. I explained this to her husband and told him that the most I could do for her was to bring her into a place of feeling loved, and to temporarily ease some of her pain.

He explained this to her and she seemed very pleased at my words. I began the healing energies and concentrated on filling her body with the energies of love. My purpose for the love energies was to heal the causes and distress associated with the arthritis. I treated her for thirty minutes and when she left, she was smiling and the tears in her eyes thanked me.

Her husband brought her back twice a week for four weeks. At that time, I began to feel that I was only serving as a pacifier for her. She kept feeling that I was going to cure her. She kept hoping for a miracle and often went home in a state of depression. I finally told them that I had to stop treating her. She pleaded with me to continue coming saying that she always felt better after a treatment. I consented but only under the condition that she accepts the limitations of my work with her.

Her treatments continued for four months. By that time, she had become too weak to get out of bed. Several months later her heart failed and she passed away.

The main point is to be realistic at all times. You do not serve others by giving them false hopes. That will cause more emotional damage and possibly make the disease more severe in its expression.
No one should ever be denied a healing. No matter how severe or advanced the disease is you can always apply a healing of love and caring.

We must realize that the state of mind is of prime importance in dealing with any manifestation of disease. If the attitude can become expressed in self-love and worthiness, then maybe miracles can take place.

Malignant Tumors

The word malignant comes from the result of maligning. The synonyms that relate to this expression are "To speak evil of, to defame, and to have a negative disposition of heart towards another."

The Universal Law states the following without exception. "As you judge yourself or others, so shall the judgment return to you increased tenfold." This statement is irrefutable. In all my years of healing, I have never encountered a case of malignancy/cancer where this "Law of Judgment" has not played a major, active role in the expression of the disease.

The energies of judgment are expressed in one of the following ways:

- Prolonged anger against yourself or others.
- Destructive comparisons and expectations of yourself and to others.
- Repeated statements of unworthiness.
- The perfectionist syndrome.
- The expression of hate.
- Self-sabotage resulting from failure orientation.

In order to counsel people afflicted with cancer, it is necessary to understand the expression of life as it occurs on our planet Earth. All souls are created in the image of God. They are composed of energies that are masculine, feminine, and love vibrations. Therefore, all souls are equally worthy in the eyes of God. When a soul incarnates on the planet Earth, it places itself in a physical body of density to experience all the life energies and patterns that compose the symphony of life on this planet.

If we can accept these words, we begin to understand that the physical body is the Temple of the Soul. It is through the body and its consciousness that the soul expresses itself and its desired achievements in life. For this reason, every cell, every organ, every part of the body and every asset **MUST BE ACCEPTED, HONORED, AND LOVED AS THEY ARE!** If not, the body, through the sub-consciousness, begins to alter its expression to conform to the conscious patterns of expression.

If you are requested to supply healing energy to a person with cancer, you will have no effect if you proceed without counseling preceding the healing. If you are not qualified to counsel, send them to someone who is experienced in that field.

Cancer is a karmically induced form of disease. Unless the person becomes completely aware of the mind conditions that have brought forth the cancer, healing cannot be effective. I cannot overstress the importance of counseling.

Over the years there have been many instances where the proper counseling has resulted in the commencement of the healing process without the assistance of energy treatments. The change in attitude and the letting go of the pattern of the probable cause of the malignancy can often trigger a remission of the tumor.

Cancer involves the destruction of healthy tissue. This places it in a category that is very different from other, normal healing applications. The energy treatment for cancer is unique in its approach and in the ultimate goal resulting from the healing session.

We throw out all the rules and learned techniques. We state that we are serving totally as a vehicle for God. We are there to transmit God's energies of transmutation to the diseased organ for the purposes of regeneration of tissue. This is the role of the healer in treating malignant tumors.

The healer opens their Crown chakra, draws in God's Light into their body, and places their hands directly on top of the tumor. With a power thought, they transmit God's Light directly from their hands into the tumor.

In this category of healing, it is of the utmost importance to maintain total objectivity in your emotions. You cannot allow yourself to become emotionally or personally involved in this unique expression of healing. If you do become involved, you will be judgmental, and YOU will be trying to cure the person. You cannot accomplish this alone. You are not God, and you cannot decide who should be cured and who should remain with disease.

In this healing process, you are not concerned with removing any form of infection or inflammation. You do not try to remove any energy from the physical body.

You make the statement in your mind that you are inserting energy to transmute all diseased tissue, enabling the person to return to the proper expression of health.

Place your hand on the tumor until you feel inside yourself that the healing has been completed for that time. This will rarely take longer than ten minutes. Repeat the application every three days for a series of five treatments. If no improvement results, there is not anything more you are able to do for the person pertaining to transmuting their disease.

What are you supposed to do if a person comes to you and says, "I feel a lump in my breast. I don't want to go to the doctor. I want you to heal it for me?" You are not a physician. You are a spiritual, energy healer. You do not ever diagnose any form of disease. If you do, you are practicing medicine without a license and are out of order and integrity.
When a person comes and asks for a healing, the first question you ask them is, "Have you been to a doctor?" If they say, "No", send them to a doctor. Do not heal anyone without a proper, professional diagnosis. If the person is currently under the care of a physician, do not give them a healing without the written permission of the doctor. You must not interfere with the practice of medicine.

This is particularly important when dealing with cancer, as it is a life threatening disease. You certainly do not want to put yourself in a position of being responsible for a loss of life because of your over-enthusiasm, or an over-active ego. Follow the rules. Be a healer. Let the doctor be the doctor.

Case Histories

1. A student of mine had her sister visiting her from another city. She came to see me for a healing as she had a herniated mass on the right side of her abdomen. In talking to her, she said that she had been to a physician who told her it was more of an inconvenience than harm to her body. I asked her why she had come to see me. She said that people told her that it didn't look nice.

The strange thing was that I felt that her condition did not truly bother her. She seemed perfectly happy with it. I felt that she actually derived pleasure from talking about her condition.

After an hour of talking, she asked me if I was going to give her a healing. I said, "No, I am not." I told her that I felt she was using the condition as part of her life. She began to cry, and told me that she didn't feel that people really liked her. With the condition, everyone paid attention to her and genuinely seemed concerned about her for the first time in her life.

I told her to go home and to stop using her condition as a crutch for her self-judgment. I said, "Forget about the hernia; just try being yourself and perhaps you will discover that others like you for yourself. If you realize this, tell the hernia that you do not need it any longer and to go away."

She left my office and returned home. The next year she appeared at my office again. She was full of smiles and proudly showed me her flat stomach. She told me that she took my advice and ignored her condition. Much to her surprise, everyone still seemed to like her and she began to develop some self-worth. She finally decided that she really did not need the hernia any longer. She told her body that it was time to be whole and the hernia began to shrink and finally disappeared.

2 A physician sent a woman of forty-five years of age to see me. It had been confirmed by several cancer specialists that she had nine tumors clustered along her spine. Seven of these tumors had been diagnosed as malignant and she was scheduled for surgery in ten days. She was sent to me to teach her to meditate as preparation for her life after surgery.

I asked her if there was a person or situation in her life that had caused her prolonged anger or resentment. She denied everything. I kept probing until she broke down and told me the following story.

She was divorced for ten years and needed to work to support herself and her four children. Her job provided her with a good living but she hated the man she worked for.

The word hate is the total judgment. She kept looking for another job, but could not find one that would provide her with sufficient income to support her family. She felt trapped, without any way to alleviate her situation.

We found the key! I explained to her that she did not have to like her boss, but she had no right to judge him. With these words, I meant that he had the right to be himself and act in any manner of his choice. She had to learn to honor his right to his own expression. Basically, she had to release any expectations she had relating to him, acknowledge who he said he was, and let him be that person.

I taught her to meditate, to relax, and to release all the anger and hate-energy, as it has become total judgment. I had her repeat to herself in meditation, "I do not like you, but I honor your right to be in your choice of expression."
She came to see me every day for seven days. Each day she meditated, and in the meditation released her prior judgments and anger. I actually never gave her any healing. All I did was to make her aware of the probable cause of her expression of disease, and to support her in allowing herself to be whole.

The following week she went for surgery. All they found were two benign tumors. The malignancies had undergone a total remission.

This is a classic example of the power of the mind, and its role in relation to the physical body. If I had given her a series of healings without the counseling, she never would have achieved her total well-being health. She had to understand what was her role in creating the illness. She had to learn from the results of her mental actions. And she did.

3 A thirty-five-year-old student of mine was involved in nutrition, healing, and counseling for many years. She was an extremely strong woman, and had married a man with a dependency personality. She played the role of parent and was his tower of strength for many years.

Finally she decided that the marriage could not work any longer and moved out of the home. He kept calling and becoming despondent.

She began to feel guilty and, once again, began to fill his needs. She could not find the mental strength to complete the separation.

Three months later she developed cervical cancer and decided to heal herself. She began all types of holistic cancer treatments and also came to me for energy healings. She knew what had caused the cancer — her refusal to honor herself as a woman and to allow herself to be worthy enough to have a fulfilled life in personal freedom.

Nothing was successful. I finally advised her to have the surgery. I suggested that she needed to understand that sometimes we need to have something cut out of our life. I told her that when the surgery was completed, she must consider that now she was free. She could have new beginnings and the courage to begin life in her own truth and freedom.

That was exactly what happened. Upon her recovery, she divorced her husband and three years later remarried. She learned from the results of her actions and has brought fulfillment and joy into her life.

This is an example of a healing that did not succeed. There was no process of healing that took place in the physical body. However, the mind is in control of life, and if we program ourselves properly, we can use the resulting action to create new stimuli for ourselves. It all depends on our mind, our motivation, and our true needs in life.

4   I received a call from a man whose eleven-year-old son was living with his ex-wife in another city. The child had collapsed for no apparent reason. In the hospital they found an inoperable brain tumor. The child was examined by several specialists who concurred with the original diagnosis.

The physician told the parents that there was no hope and to be prepared for their child's death. It was at this point that I received a request for a healing for the child.

I mentally sent the child healing energy. The conditions were as follows. The energy was sent to the soul of the child to allow the soul to utilize the energy according to its truth and needs. With these conditions, I do not interfere or make any decision relating to the life or death of the child. I am only making the energy available to the soul.

Three days later, the child had undergone a total remission of the tumor, awoke, smiled, and went home.

I have no explanation for what happened. I take no credit for the results of the remission of the tumor or his complete recovery. It was truly in the hands of his soul and God. I have attempted healings of this nature many times. They are not always successful because of many factors. All I can do is to be there when called upon. I am not the final judge or deciding factor in the process of curing of any expression of disease. Actually, I would not want to have that great responsibility for life or death.

In every case I have described to you, as well as the hundreds of others, the individual has remained under the care of their physician. Those who were being treated with radiation, etc., continued their treatments. This is most important for you to understand. A healer has no right to say, "I will make you well. Stop your other treatments." That type of action would be irresponsible and out of pure ego.

Two of my dear friends developed cancer within three months of each other. She had lymphatic and bone marrow cancer. Her doctor did not have any hope for her survival. He developed two brain tumors.

One year later, she was completely well and was released by her physician. She is still well fifteen years later. He passed away in two months.

Both of them came to me for healings in addition to continuing medical treatment with the approval of their physicians. She was able to recognize and release much past anger that she had held onto for fifteen years. He was unable to forgive and to express his true feelings. He always told people what they wanted to hear to get their approval. Maybe she recovered as a result of her medical treatments. It does not matter. I don't need the credit or the accolades. She recovered; that is what matters. If the healing only had a psychological effect on her, it was worth the effort.

A healer has no right to judge themselves if a person does not recover from disease. The role of the healer is limited to supplying energy. Anything else is between the individual and the Creator-God.

If you cannot be compatible with these words, I advise you not to enter the field of healing. All that will result for you will be frustration and self-judgment when you are not successful in your application of energy. We are not here to question how and why? We are here to serve. The complexity of the universe and the laws of God are beyond our comprehension.

Benign Cysts and Tumors

These expressions of what I call a lack of well-being are often considered to serve as a warning system to the person. It is like being tapped on the shoulder and being told, pay attention, you are doing something or thinking something that is out of truth and order.

Many people ignore benign cysts or tumors feeling that they will just go away. Most times they remain and if ignored can develop into malignancies. Ignoring the existence of anything physical is like living in illusion.

The reason for its existence will continue to grow in power, as the cause becomes increasingly stronger in the conscious expression. This could lead to their development into malignant conditions.

The process for healing and treating benign tumors and cysts is the same procedure as described in healing malignant tumors. Even though the condition is benign, we treat it as if it were malignant as a precautionary measure. Counseling is required, and the cause of the disorder must be determined before the healing can take place and be effective.
Parkinson's, Hodgkin 's disease and Multiple Sclerosis

We have classified these three expressions together, as they are connected to part of a group of common causative factors that result in classifying disease.

The dominant spiritual causes for these expressions are prolonged anger, resentment against others, the need for attentive love, and self-pity.

Case Example

A chiropractor friend of mine called me to tell me that his forty–five–year–old father had developed Parkinson's disease. He asked me to speak to his father and to assist him if possible. I went to see the man. He had been diagnosed with this disease eight months earlier and was already confined to a wheelchair. This was a very rapid progression of the disease.

This was the man's story. He worked for an international chain of motels and was promoted to national sales manager. He tried to refuse the promotion, as it would involve extensive travel each month. Even though he had worked for the organization for fifteen years, they told him that he could not refuse the job. Either he takes the job or leaves the company.

He accepted the position but was very angry at being forced to alter his style of life against his wishes. When he was away on business he drank very heavily, and vented his anger on people around him. Finally he became ill. The illness increased his bitterness and anger.

He totally blamed the company for his condition and refused to even consider any degree of self-responsibility. He would not listen to my words and openly refused to let go of his anger. Within one year the man had died. His progressive deterioration was so rapid that even his physician could not comprehend what had taken place.

The disruption of the nervous and energy systems is closely related to the spiritual results of prolonged anger and resentment. If you have a fireplace and do not keep putting logs on the fire, soon it will cool down and go out. If you keep feeding it fuel, eventually your house will burn down.

The angrier you become, the more you cause a disruption in your stability and state of mind. This results in creating a stronger susceptibility to energy and nerve disorders.

In the final analysis, I have not found a condition of life where there is no choice. This makes anger at, or judgment of others invalid. The man could have quit his job. It would not have been the end of his world.

There are always choices that can be made, even if they involve a temporary step backwards.

The man's anger was misdirected. His anger was at himself. He accepted the job, and he alone was responsible for his action and for the results. Therefore, we can say that his anger and rage were directed to himself. The aggression was expressed to his company.

A.I.D.S

On the day that the world became aware of the existence and dangers of A.I.D.S. a hand "appeared" and closed the curtains on the visible window of promiscuous sexual activity.

The medical community has associated A.I.D.S. with a breakdown of the body's immune system. This is in total accordance with spiritual law. The spiritual interpretation of the immune system relates to the balance and alignment of the chakras, the spiritual energy centers through which the life energy flow to all areas and centers of the body.

As long as your body energies are flowing unobstructed, your body should not be subject to any form of disease. Every judgment, every jealousy, every action that is a lie, not truth, causes distortion in your energy flow. This results in a shortage of energy going to specifically related areas making them susceptible to disease.

Do you ever ask yourself any questions before you decide to be intimate with someone? Such as, "Am I taking this action in truth or in a lie? Do I want to be promiscuous or do I need to prove something to myself? Do I feel undesirable and need my ego boosted to feel good about myself?"

These are some of the questions I pose to anyone who comes to me for healing of the condition of A.I.D.S. The very first question I ask is, "Tell me, do you want to live, or to die?" If they want to live — and most do — they must begin to face all their past actions, and face them in truth. They cannot lie to themselves any longer. The life of illusion must end.

The mind must force the ego expression to the side and allow the soul truth to become the dominant expression that creates all future actions on the part of the individual.

I truly believe that sometimes A.I.D.S. can be healed through the initial process of counseling and creating an awareness of past actions that were not taken in truth. At this time, I must state that our success is quite small, but remission has taken place in some cases. If it works only one time, it can work again and again.

When I ask someone if they wish to live, I am asking them if they are prepared to stop lying to themselves and to honor their body as the temple of their soul. This is not an easy task. They need a support system to encourage them and to show them that someone truly cares if they live or die. In other words, do they need the expression of love and nurturing instead of just passing sex, for there is a vast difference between the choices?

Every time a person with A.I.D.S. acknowledges an untrue action of the past as a lie, the chakra system moves a little closer to being balanced. This is the true healing process. Step by step, move into truth and bring the immune system into alignment. If this disease can be treated, in time perhaps the person can be returned to health.

I have not said that I cure A.I.D.S. I don't. This is an expression of disorder that must be cured from within the individual if it has a spiritual energy cause and expression. If the person coming to you with A.I.D.S. begins the path of self-exploration, you can proceed with energy healing. Place a hand on their heart chakra and send love right into their heart and soul. Having the experience of love without any demands or attachments will allow them to begin to have feelings of worthiness and value. These can be just the right incentives for them to want to heal, and want to be well.

Congenital Disease

There are many medically proven causes for congenital expressions of disease and we are not going to disagree with any of them.

Science has made wonderful advances and opened many doors in understanding prenatal conditions of disability. Our discussion is confined to the spiritual or soul causes for an expression of disability, whether it be of physical or mental nature.

One of the basic precepts of spiritual philosophy is that the soul chooses the parents. In this manner, the soul will receive the desired genetic, hereditary, and environmental conditions to expose it to the experiences it has chosen to have during that lifetime. It has also been acknowledged by the medical community that the soul enters the fetus at the time of conception. This normally insures the complete and healthy formation of the child.

It is said that, at times, the soul chooses an incarnation with a disability, either to serve others or for its own experiences as a minority part of a society. When this occurs, the soul does not enter the fetus at conception but allows time to elapse, creating the desired expression of disability. Then it enters the fetus.

This statement may be difficult for you to comprehend and to accept. You might ask why a soul would choose a lifetime being physically handicapped or mentally disabled. I shall do my best to explain this concept to you.

In a spiritual sense, if a child is born with a mental or physical defect, in a minority color or race, any condition of expression that is not the majority of society, we call it a minority expression.

All types and expressions of this category of disease are usually for the duration of the lifetime. Therefore, any type of normal healing procedure that could result in a possible cure cannot be effectively applied. The healing process that should be utilized comes from the power of your mind, the mental projection of energy to the person's soul.

There are many reasons for this.
1. It serves as recognition of the presence of the soul, as well as acknowledging its existence.
2. You are acknowledging that it is the soul's lifetime, not just the personality.

3. You are making energy available to the soul, to assist it in expressing and completing the purposes for which it elected that expression of life.
4. The soul will become aware that it is not alone, locked in a disabled body, and that someone else truly cares and supports its choice of life.

When you meet someone in a minority expression, do not feel pity or any other type of judgment. Go behind who they seem to be. Look inside, feel who they really are, and allow yourself to relate to their core of life, their soul.

It has been my experience that people living with a disability from birth have strong, evolved souls. They have chosen their life, a life of restricted expression, and yet are here to experience, learn, and serve others. The majority of these people are gentle and loving. This expression comes forth from their soul to touch people and teach them the true value of life.

Spiritual Causes of Disease

The human body is susceptible to expressions of disease. This is a reality that everyone must accept. Are these words true, or are we capable of believing the bibles of the world that tell us people lived for hundreds of years?

Have we lost something along the way? Was it our industrial revolution, our capitalistic societies, our drive for success and power? Maybe it was all of them and much more.

Physical disease is real and cannot be denied or ignored. However, are there other causes and reasons that make us susceptible to expressions of disease?

With few exceptions, our bodies are created in perfection and, according to Spiritual Law, should remain that way. When our chakra system is in order and alignment, our immune system maintains our well-being and denies the presence of imperfection, or disease.

During the last twenty years, I have worked with many hundreds of people that have contracted various forms of illness. I have rarely encountered anyone whose illness did not have a spiritual cause that triggered the inception of their disease. The spiritual cause re-programs the energy response centers of the body and makes them receptive to the mind-statements that deny the perfection of our creation.

We are susceptible to disease but, if our immune system is in order, the severity of the illness is not so intense and the duration of the confinement could be greatly shortened.
We can compare our sub-conscious mind to the information stored in the memory banks of a computer. When we, from the thoughts of our mind, begin to create a new program, the repetition eventually cancels the original program.

The body cannot distinguish truth from untruth. It just listens to the subconscious automatic response system and can begin to program itself for an expression of disease. The body never makes the mind a liar!

During the course of listening to people's problems over the years, patterns of programmed causes of disease have created categories of negative thought statements. These can alter the reactions put forth by the subconscious mind and have a serious effect on the healthy energy of the related organs of the body, making them susceptible to disease.

In this section of the book, I have categorized the negative statements of affirmations that become transmitted to the organ energy of the body, as well as to the chakra centers that relate to that organ. I have also listed the positive affirmations that can be used to negate the past expression and begin to recreate the positive flow of eternal health to the structure.

I have listed many expressions of disease. The ones mentioned I believe to be the most common conditions we face in our lives.

Acne

- No matter what people say, I know that I am not worthy.
- I wish I could have plastic surgery to change my ugly face.

    Affirmations of change:
- I celebrate my imperfections and mistakes.
- My true self comes from my soul.

Adrenal Glands

- No matter what I try, I am doomed to failure.
- I don't know how to relax.

    Affirmations of change:
- All my actions of life are successful.

A.I.D.S.

- I am a worthless human being.
- I place no value on sex or love.
- I constantly need to prove I am desirable.

Affirmations of change:
- I am perfect for my soul and the Creator-God.
- I accept and love myself as I am today.

Allergies

- If I try something new, I know it will not work.
- I don't do well under pressure.
- I'm not good enough to handle success.
- I feel insecure with people I don't know.

Affirmations of change:
- Constant change is a joy in my life.
- I thrive on new adventures.
- I accept myself when I look in the mirror of my reflection.

Alzheimer's Disease

- Life is too difficult for me.
- I can't seem to relate to anyone.
- I wish I were born in a different generation.

Affirmations of change:
- I accept myself and all others in Grace.
- My soul is my personal treasure.

Amnesia

- I wish I could forget everything that has happened to me.
- Please don't remind me of the past.
- If I could only live on a deserted island.
- I can never defend myself; my throat closes up.

Affirmations of change:
- I accept the reality of life
- Today is the first day in my life

Anemia

- I always stay on the fence instead of making decisions.
- I am afraid of life and new circumstances.

Affirmations of change:
- I celebrate my mistakes and learn from them.

Anorexia

- I hate my body.
- I am rejected by everyone
- I am afraid to accept anything in my life.
- I am not worthy of receiving anything.

Affirmations of change:
- My soul never rejects me.
- I am perfect for my soul.
- I am worthy of receiving abundance.

Appendicitis

- I never allow myself to receive good things.
- I am afraid to visit new places.
- I always have trouble making new friends.

Affirmations of change:
- I welcome the flow of the abundance of life.
- My body welcomes all changes in life.

Arthritis

- I am not a good person.
- I am a failure.
- I am not worthy for them to love me.
- I harbor anger and resentment towards others.

- I don't want to forgive them. I hate them.
- I am a perfectionist and will accept nothing else.
- I expect everyone to "do as I do" and "be as I am".

  Affirmations of change:
- I accept success with joy.
- I accept all other people in Grace.
- I celebrate all my mistakes.
- I have no expectations from anyone.
- I am an imperfect human being.

Asthma

- I was told never to cry.
- Expression of emotions is a sign of weakness.
- Love is only for those who live in illusion, not reality.
- I am always blocked whenever I try to accomplish anything.

  Affirmations of change:
- My emotions are my personal power.
- I am in the flow of prosperity and success.

Bladder

- I allow other people to dump on me.
- I know that everyone knows better than I do.
- I always do things to please others at my own expense.

  Affirmations of change:
- I am responsible only for my life.
- I accept myself as I am today.
- Sacrifice is not a part of my life.

Bowel problems

- I want to hold on to what I have and know.
- I never seem to get enough to satisfy myself.
- I really don't want anyone to find out who and what I am.

Affirmations of change:
- I accept all new experiences in life without fear.
- I welcome the flow of change in my body.

Breast Cancer

- It must be my fault. I am not a worthy mother.
- He doesn't want to make love to me. I am a failure as a wife.
- My breasts are too large. I hate them.
- I'm flat-chested. I don't even feel like a woman.
- I wish I were a man.
- I had an affair. I feel guilty and deserve to be punished.

Affirmations of change:
- I am responsible only for myself.
- My body is perfect for my soul.
- I accept myself as a woman.
- I accept the truth of others.
- I learn from all experiences of life without self-judgment.

Bronchitis

- I just can't seem to express myself clearly.
- I have trouble relating to people.
- I don't want to fit into society.

Affirmations of change:
- I always speak clearly.
- I relate easily to all people.
- I am myself at all times.

Bursitis

- I always want everyone to like me.
- I play roles in life to please others.
- I avoid expressing anger. I hold it inside.

Affirmations of change:
- I enjoy communicating with people and speaking my truth.
- I enjoy just being myself.

Cancer

- I always compare myself with others and feel inferior.
- I harbor anger and resentment to others.
- I always hold on to guilt and punish myself.
- I am not a worthy person.
- I hate my body.
- I always judge myself and everyone else.
- No one will ever love me. I don't even love myself.

  Affirmations of change:
- I celebrate my uniqueness in life.
- I am perfect for my soul.
- I accept all people as they say they are.
- I am in Grace to myself and all others.
- I accept and love my body with all its imperfections.

Cerebral Palsy

- I give up with everything.
- I can't cope with normal society.
- My family are total strangers and don't understand me.
- I feel all-alone even at the deepest levels inside of myself.

  Affirmations of change:
- God and my soul never reject me.
- I am a shining star in life.
- I accept my uniqueness in life.

Colitis

- I am always sacrificing myself for others.
- I always feel guilty when I fail in something.
- I have trouble making decisions.

  Affirmations of change:
- Sacrifice is not a part of my life.
- All my actions are my truth.
- I always make clear decisions.

Diabetes

- I expect to be rejected by most people I meet.
- I don't feel comfortable in most situations.
- I have trouble getting people to love me.

  Affirmations of change:
- I accept myself in Grace as I am today.
- I am complete in my love for myself.

Emphysema

- I always struggle for survival.
- Nothing ever comes easy for me.
- I feel closed in when I am in unfamiliar surroundings.

  Affirmations of change:
- I breathe in the flow of life.
- Peace and joy are the expressions of my life.

Epilepsy

- I feel that I am on overload all the time.
- I have difficulty committing myself to something.
- I always want to make the right decision.
- I am always afraid of becoming insecure.

  Affirmations of change:
- I deal easily with all stress.
- I celebrate and learn from all my mistakes.

Glaucoma, Cataracts, Corneal Conditions

- I don't want to see what is out there.
- Don't make me look at what causes sadness and despair.
- If I don't look, maybe it will go away.
- I don't want to relate to life.

  Affirmations of change:
- I accept the existence of all things I see.

- I enjoy the challenges of life.
- I accept change in my life.

Gall Bladder

- I always listen to other people, not myself.
- Everyone usually knows better than I do.
- I always let people impose themselves on me.

   Affirmations of change:
- "No thank you" are my favorite words.
- I accept responsibility for my life.

Genital Organs

- I wish I didn't have any genitals.
- No one truly loves me.
- I am not supposed to talk about "that".
- I wish I were a woman.
- I wish I were a man.
- Sex is evil and dirty.
- I am ashamed of my body.
- I have sex only to use and control someone.

   Affirmations of change:
- My genitals are one of my body's treasures.
- I am perfect for my soul.
- My body is the tool of expression for my soul.
- I am proud of my body.

Hearing Problems

- If I don't pay attention maybe it will go away.
- I don't like surprises.
- Don't say anything to change my way of thinking.
- I can't stand it when people place doubts in my mind.
- I avoid disagreement and conflict.

Affirmations of change:
- I accept the existence of all people and things without judgment.
- I enjoy defending my truth.
- I like hearing different opinions.

Heart Disease

- I don't believe in God.
- There is no such thing as a soul.
- I always keep a wall around my heart so no one can find out how unworthy I am.
- I have never felt love in my entire life.
- I don't know how to be happy.
- I never cry.
- I don't like people to touch or hug me.

Affirmations of change:
- I am a child of God.
- God loves me.
- My soul loves me.
- I allow joy into my life.

Hepatitis, Liver

- I sacrifice my own needs to do for others.
- I allow other people to impose their will on me.
- I do not take time to discover who I am.
- My self-esteem is low.
- I make emotional decisions without thinking.

Affirmations of change:
- I take all actions in my truth.
- I am responsible only to and for my truth.
- My body rejects all energies of sacrifice.

Herpes

- Every time I have sex, I feel dirty.
- I have no real value of my body or myself.
- I always let people use me. What's the difference?
- Who needs love? I just have fun.

   Affirmations of change:
- I love and value my body as a creation of God.
- My body expresses love.

High or Low Blood Pressure

- I never take the time to determine how I feel in most situations.
- I am not a peaceful person inside.
- I don't know how to relax. It makes me feel guilty.
- I thrive on tension and stress.

   Affirmations of change:
- My body is always free of tension and stress
- I enjoy the art of relaxation.
- I never make quick decisions.

Hodgkin's Disease

- I always bend over backwards to please people.
- My life is one compromise after another.
- I never seem to enjoy things just for myself.

   Affirmations of change:
- I am the most important person in my life.
- I enjoy actions of self-accomplishment and success.

Hyper Activity

- I love to have ten things to do at the same time.
- I never have time just for me.
- Anything less than perfection is failure.
- If you need a favor, just ask. I love to put myself out for people.

Affirmations of change:
- I take the time for self-discovery
- I always learn from my mistakes.
- Being imperfect creates freedom of expression.

Hypoglycemia

- I never feel balanced.
- My world is always upside down.
- I never get back what I put out.
- I always feel cheated in relationships.
- It is always an effort for me to complete things.

Affirmations of change:
- I am in control of my realities.
- I am motivated to be successful.
- I create balance in my life's situations.

Kidneys

- I never take time to validate actions for myself.
- I am always in a great rush with my life.
- I allow others to impose their will over me.
- I never feel worthy of having success.
- Most people know better than I do.

Affirmations of change:
- I make all my decisions in my truth.
- I am always successful in my actions.
- I take full responsibility for my expression of life.

Leukemia

- I always have a defeatist attitude towards myself.
- I have no incentive to follow through on my actions.
- When I face any obstacle, I just collapse.

Affirmations of change:
- I accept my imperfections in Grace.
- I am unique as the expression of my soul.

- God loves me as I am.

Lymphatic Disorders

- I do not ever honor my body.
- I place the value of others ahead of myself.
- I take responsibility for everyone.
- I allow other people's problems to affect my life.
- I worry about people I love all the time.

Affirmations of change:
- I love and honor my body as the house of my soul.
- I am responsible only for my life.
- I am the most important person in my life.

Mental Depression

- My perception of success is considered failure by others.
- I keep proving to myself how unworthy I am.
- I attract losers into my life.
- Whatever I touch becomes a disaster.
- Why bother? I know how it will end.

Affirmations of change:
- I am a wealthy person.
- My soul loves me as I am.
- I live in the flow of abundance without limitations.
- I am worthy of success.

Migraines

- I play mental games with people.
- I enjoy hiding my true self from others.
- I avoid situations with unknown outcomes.
- Pain reminds me of my inability to achieve success.
- When I suffer, people are drawn to me.

Affirmations of change:
- I always celebrate when I make a mistake.
- I enjoy not being perfect.

- I always relate to people in my truth.

Multiple Sclerosis

- I harbor anger and resentment to others.
- I am not a worthy person.
- I always hold on to guilt and punish myself.
- When something goes wrong, I know it must be my fault.
- I judge myself and other people.
- I am always a failure.
- No one can love me. I hate myself.

   Affirmations of change:
- I accept myself and all other people in Grace.
- I accept all my imperfections.
- I speak my truth at all times.
- My soul and God love me. I am a wealthy person.

Muscular Dystrophy

- I am tired of failure. Don't try to change me.
- My body rejects the power of expression.
- I always compare myself to others.
- I always have expectations from others and become disillusioned.
- No matter what happens, I always go back for more.
- I am really angry at the world.
- I look around and resent the success of others.

   Affirmations of change:
- My body expresses itself in truth.
- I am the mirror of my soul.
- I live in freedom, peace and joy.
- I carry the sword of courage at all times.
- I love my body and my strong mind.

Myopia

- I always lead my life in a tunnel.
- I don't want to think of the future. It makes me nervous.
- I am uncomfortable dealing with the unknown.

Affirmations of change:
- I anticipate the future with joy.
- Challenges nourish my mind.
- I am responsible only to myself.

Neuromuscular Problems

- I am insecure dealing with situations whose outcome is insecure.
- I don't know how to handle success.
- I have a very weak mind.
- I cannot concentrate on anything.
- When I fail, I want to curl up in a ball and hide.
- I always feel angry with myself for everything.

Affirmations of change:
- My existence is the power of my mind.
- I accept myself today in Grace.
- I am nourished by changes in life.
- I am secure in the love of my soul.

Obesity

- If I shield myself, no one will discover who I am.
- I need protection so I will never be hurt or rejected again.
- I don't want to discover who I am. What if I don't like the results?
- If I let out the real me, I might be tempted to fool around.
- I am safe hiding from other people and myself.

Affirmations of change:
- I really love myself as I am.
- I am perfect for my soul.
- I am my true self all the time.
- I celebrate my imperfections in life.

Osteomyelitis

- I am all alone, without support from anyone.
- Life is full of frustrations for me.
- Why was I ever born?

Affirmations of change:
- My soul is my best friend.
- I am a creative and independent person.
- I enjoy walking through life.

Pancreas

- I have trouble accepting the reality of my own truth.
- I doubt my ability to make decisions.
- I am a very gullible person.
- I hate being in the middle. I vacillate and close down.

Affirmation of change:
- I value the decisions of my mind.
- I am worthy of my inner realities.
- Making decisions gives me power.

Parkinson's Disease

- I will never forgive them for what they have done to me.
- It is their fault that I failed.
- Everyone is out to get me.
- I hold anger inside myself so that others will accept me.
- No one has really shown me love in my whole life.
- Everyone has always told me that I am a failure.

Affirmations of change:
- I accept all people in Grace.
- I take responsibility for my actions and my life.
- I live in the present. Today is my life.
- I am so wealthy. I have the love of my soul.

Pituitary Conditions

- I have always been called stupid.
- Whenever I face the unknown, my mind turns into a bowl of jelly.
- Sometimes I think my mind is on permanent vacation.
- I have absolutely no self-control.
- Anyone can convince me of anything.

- I don't believe in God.

    Affirmations of change:
- I celebrate the uniqueness of my soul.
- I live in the abundance of life.
- I am a child of the Creator-God.
- My mind is the tool of my expression.

Polio

- I always compare myself to others and come up short-changed.
- I have been overly possessive all my life.
- I don't like to share my things with others.
- I sabotage others and enjoy their failure.

    Affirmations of change:
- I move forward into life's expression with anticipation.
- ØI am unique in my expression.
- ØI always share the joys of my life.

Prostate Conditions

- I'm ashamed of what I have. It's much smaller than the other men.
- I avoid making love from shame.
- Who needs sex? I enjoy a good conversation.
- I'm too old for such things.
- At my age, I'm not supposed to think about sex.
- I became celibate as a sacrifice to God.

    Affirmations of change:
- I honor the functions of my body.
- My body is a tool for sharing truth and love.
- God accepts me just as I am.
- Life is the full expression of love.

Psoriasis

- I like others to be responsible for the results of my actions.
- If I take chances, I always get hurt.

- I try not to express my true feelings. I just say what pleases people.

   Affirmations of change:
- I am in control of my life.
- I celebrate my imperfections.
- I enjoy the constant flow of change in my life.

Rheumatism

- No one in my family expresses affection.
- I have always been blamed for everything.
- I don't know, but I am always angry.
- Love is something I avoid talking about.

   Affirmations of change:
- I accept love in my body.
- My body nourishes my soul.
- I walk joyfully into the adventures of life.

Sciatica

- I am always afraid to enter new ventures.
- I always wear everyone else's problems.
- I am always safe by just repeating my past experiences.

   Affirmations of change:
- I am excited by the unknown.
- I am responsible only for myself.
- New experiences contribute to my life.

Sinus Conditions

- My mind gets confused with the slightest pressure.
- I have trouble maintaining clarity of thought.
- I do not function well under stress.
- I let people irritate myself.

   Affirmations of change:
- I thrive with the pressures of daily life.

- I am always in control of my mind and thoughts.

Skin Conditions

- I have anxieties and fears relating to my life.
- People always are a threat to me.

   Affirmations of change:
- I accept life as it unfolds for me, day by day.

Spine

- I do not feel supported in life.
- I mistrust everyone.
- I take on the weight of the world.
- I am never worthy of real prosperity in life.

   Affirmations of change:
- I am in Grace to the world.
- I am supported in my choices of life.
- I accept people as they are.

Spinal Meningitis

- Everything in my life is chaos.
- I hold onto old ideals and patterns of life.
- I can never find out who I am.

   Affirmations of change:
- I anticipate the joy of change.
- I am the expression of my soul.
- I accept the balance of expression in my daily life.

Spleen

- I have a strong drive for perfection.
- I have an addictive personality.
- I become obsessed with things that I enjoy.

Affirmations of change:
- I accept variety in my life.
- Being imperfect relaxes me.

Thyroid

- My whole life is one frustration.
- No one ever listens to my needs.
- Life is too complicated for me.

Affirmations of change:
- Life is simple joy.
- I always express what my truth is at that time.

Tuberculosis

- I always seek revenge against others.
- My dreams are full of hate and cruelty.
- I am not a forgiving person.
- I want my pound of flesh.

Affirmations of change:
- Grace is my password in life.
- I love myself and my soul.

Ulcers

- I never say what is on my mind.
- I hold back my truth so as not to make waves.
- I usually feel inferior to others.

Affirmations of change:
- I confront all conflicts in my life.
- I always speak my truth to others.

Venereal Disease

- I feel guilty whenever I enjoy sex.
- I am ashamed of my genitals.
- I use sex for power and conquest.

Affirmations of change:
- I honor all parts of my body.
- I accept myself as a complete man/woman.
- I express my emotional feelings in my truth.

I wish to reaffirm to you that most of the statements relating to disease are reactive patterns from the subconscious mind. They are created from conscious thoughts resulting from actions taken or not taken in life.

Grace

It is necessary for me at this point, to define the word Grace in my frame of reference and usage. People speak of reaching a state of unconditional love. It sounds quite romantic. That is the problem. In our societies, we relate the word love to our emotional and sexual expressions. That is the nature of our lives here. Emotions and sexuality are karmic experience vibrations that are part of the evolutionary experiences of this planet. They do not exist in the universal vibrations of permanence. Therefore, we cannot reach this pure state in a physical life here on Earth.

We can, however, place ourselves in two expressions: A state of Grace to all parts of ourselves as well as to other people.

Grace means the following to me:
This is the acceptance of all aspects and expressions of yourself as you are today without judgment. That does not mean that you have to be satisfied with yourself or the place you are in life. You have the right not to like certain things about yourself. The point is this. Today, this is who you are. Accept it and then change if you want to. You cannot change anything you do not acknowledge exists. Denial prevents change.

Stand naked in front of your mirror. What do you see? Who you are today! That is all you have. You don't like it? Change it, but don't deny what you see. Then you are in Grace to yourself.

Grace to other people means: I don't have to like you or associate with you but I acknowledge your right to be whomever you choose to be without judgment.

If you can achieve this state of acceptance, your whole life will change for you. You will become free of everyone and allow all people to have the complete responsibility for their own lives.

The great majority of statements that are part of the causes of disease originate from the programming without self-Grace. The anger, judgments, guilt, sacrifice, resentments are all expressions that are out of Grace to you, as well as to other people.

If there is such an expression as unconditional love here on Earth, it is Grace to ourselves and to all people.

Universal Healing Laws

The ability to bring healing energies into your body and transmit them to others is indeed a gift from God. At some level, an unwritten agreement exists between you and God. This agreement relates to your purposes and intentions to use the energies only for the benefit and truth of others in total integrity.

Many years ago during my own early development, I was told by spirit that anyone who uses the healing energies for personal gain, to control others with power, or for greed, will lose the privilege and availability of the healing energies of God.

Several months later during the course of the day, words came into my mind. "The ability to generate and transmit healing energies through yourself to others is a sacred trust bestowed upon you by God. If they are used without the will of God, the power will be removed from your being."

I thought about these words for a while and decided that spirit was only trying to test me. My ego said, "Once I have this power, it will not leave. It will always be mine."

As the time elapsed, I forgot about those words. One day I was healing someone and feeling all the wonderful, warm energy flowing from my hands into their body. All of a sudden the flow stopped. It felt like a faucet had been shut off. Nothing was there. Nothing. No matter how hard I tried to concentrate, I could not bring forth the flow of energy. After several minutes I became very concerned.
Suddenly I heard some words in my mind, "The next time you are told one of God's laws, I advise you to heed the words." The flow of energy resumed at once and I was able to complete the healing. I had learned and experienced the truth to the statement, "God grants and God has the power to remove."
In the years of developing as a healer, I did not have anyone to instruct me. I made every mistake imaginable but I learned.

The result has enabled me to share the Universal Healing Laws with you as they have been related to me over many years and under a variety of different circumstances.

Because of the variability of life, there are many exceptions to healing law. Universal law is flexible and does adapt to the conditions of free-will and choice. However, we must be responsible for the results of our actions.

1.   "It is never proper, without any exceptions, to heal someone with your own, personal energies. If you do, you will draw to yourself the karma of self-sacrifice and the absence of self-worth."

We have been provided with spiritual energy receptacles in our body that are called chakras. Whenever you wish to begin a healing process, ask God to send the healing energies into your crown chakra. This center is located in the front, center of your head just above the hairline.

Remember, you are the vehicle. God is the initiator and supplier of all healing energy. If you were to use your own energy to heal with, you would become drained of your energy and could become susceptible to the other person's disease.

After you have completed a healing, if you are tired or feel drained, your procedure has not been conducted properly. This will be an indicator telling you that you have used your own energies during the healing process. You will need to re-examine your techniques and make the corrections in your procedure. When you conduct your healings with God's Light, you will never become weary or tired. As this energy flows through you, it will constantly refresh you and your own energy stability.

I will always remember my first attempt at healing. A friend of mine had developed pneumonia and asked me to give him a healing. I hurried over to try out my newly found talents. At that time, I was unaware of any healing laws or procedures. I began to send energy into his body and attempted to pull out the heat and the infection. When I was finished, I went home feeling quite elated, hoping that the healing would be successful.

The next morning my friend called me, and excitedly praised my work. His fever was gone, his breathing and lungs seemed to be functioning normally. I had to whisper my thanks for his call, as I had almost no voice. During the night I had developed bronchitis and was quite ill.

That was the first and last time I ever used my own energies to heal with!

Example

There is a well-known energy healer in America who has been a practitioner for many years. I met him in 1981 and was amazed by what I saw. The man was fifty years old and had just recovered from heart by-pass surgery. He was also suffering from chronic arthritis of the spine, the hips, and the knees.

I could not comprehend his condition until I observed his healing techniques. I stayed with him for a full day and I watched him, time after time, taking his client's illness directly into his own body. When he felt the client was clear, he attempted to transmute and disperse the diseased energy from his own body.
I questioned him about his reasons for doing this and his response shocked me. He said, "This is the only way I can be sure that the person is free of disease. I always heal this way."

What could I say? I just smiled but inside I now knew why this man was so ill. A residue of the diseased energy remained inside of him after every healing. This eventually built up and created his own susceptibility to disease. The unconscious message he created was, "I sacrifice my well-being for other people." This generates the karmic vibration of self-judgment.

Some people call this faith healing. I knew that he did not have enough faith in himself as a vehicle to transmit God's Light. He needed to experience the disease to convince himself of his ability to heal.

In the final analysis, do we have the right to attempt to take a disease out of someone's body? The answer is firm.

The integrity of healing requires that you remain detached at all times. The healer is an instrument for energy, nothing more. You are not there to make decisions on behalf of the client but to empower them to use the healing energies themselves and to be in support of their decisions.

2. "God's state of Grace requires that you never volunteer your healing services or heal anyone directly without their permission."

One of the prime purposes for this law is to relieve the healer of the responsibility of having to decide who is entitled and who is not entitled to a healing.

The Universal Law of Cause and Effect creates the circumstances that require everyone to ask for a healing. By asking, they are really saying, "I believe in God's ability to cure me and I no longer have any need for disease."

Because of the Law of Cause and Effect, the moment someone requests a healing, it begins to take effect. The statement of request is the expression of the person's truth and the return of healing energy begins to flow to them.

There are specific exceptions to this healing law:

- All members and relatives of your immediate family.
  Members of a family are usually a part of what is called a group soul expression. These souls often incarnate together and agree to become involved in common patterns of conscious growth. Because of this, you are permitted to volunteer your healing services to maintain the flow and balance in the family energies.

- Children who are too young to be responsible for their actions and their decisions created as a result of their actions.

In these circumstances, you must be asked for a healing by the child's parents, guardians, or those responsible for their welfare.

- People who are too ill or elderly to be responsible for their conscious decisions.

Under these conditions, the parties who are responsible for their care can make the request for healing on their behalf. (The procedures for this type of treatment will be discussed at length in the section on absentee healing).

I have seen healers approach someone and say: "I feel that something is wrong with you. I'll fix it for you." And then commence putting their hands on the person trying to effect a healing.

No one has the right to this type of action. We all want to be of help if we can but we must remember that it is everyone's right to be where they have chosen to be without any interference of judgment. This is difficult at times. We want to assist people but we must remain detached, objective, and just support people in their own choice of life's experience. The law must be respected and observed at all times.

If you impose yourself on someone, you take the chance of trying to heal someone who may not believe in healing. They just may be too embarrassed to refuse your offer. If this situation occurs, their body will automatically reject the energy. The body will remain in the expression dictated and desired by the conscious mind. In addition to this, you might generate fear in them. This could generate negative energy that will be directed to you and affect you in an adverse manner.

The following set of circumstances could take place. An adult member of your family becomes ill. They know that they should request a healing but say nothing to you. What should you do? My suggestion would be to wait a while and observe what happens. You could offer them a healing but maybe they just need to be sick for a while. Maybe they need a little tender loving care for a few days. If you sense this, leave them alone and just love them.

Remember that you are here to serve the needs of others. Then you will have the strength not to act from your emotions despite feeling that you could possibly help them in their current condition. The average person is unaware of the laws of healing.

They do not know that they must request the healing. If this situation happens, you are permitted to speak to someone else who can relate the proper procedure to the person who is ill. By doing this, you allow the individual to make their own free-will decision, without any interference or influence on your part.

You must remember that you are dealing with cosmic light and Universal Law. These are the most sensitive and purest of energies. Any variation could cause a distortion, resulting in ineffective healing results.

3. "Never attempt to heal an illness induced by a karmic action without the determination and acceptance of the probable cause for the existing condition."

When an individual has spiritually created a condition for themselves, it cannot be relieved until they have acknowledged the probable cause and stated their need to be whole again. If you interfere in this type of situation, you place yourself in the position of judging the person and their past actions.

The body always listens to the input of the mind. If the mind thought is an expression of unworthiness and the thought is reinforced through repetition, the body creates that specific unworthiness to maintain alignment with the mind thought. This generates the karmic disease and maintains it for as long as the thought patterns remain in consciousness.

I must caution that many times a conscious self-judgment may be forgotten. This can be dangerous. The thought energy can easily transfer to unconscious levels and remain there as a shadow of expression. If this occurs, the disease will intensify and the pattern of behavior will begin to affect other aspects of the personality expression.

If you attempt to just give a healing without any conscious understanding and acceptance, the healing cannot be successful. The body will reject the energy as a form of invasion into its programmed truth expression.

4. "Do not attempt to heal anyone with God's energies if they do not believe in God's existence as their Creator."

If you attempt to heal someone who has said that they do not believe in God, their body will reject the energies. Their chakras, or spiritual energy centers, will not be open. The entry of God's Light and essence will not be possible to achieve. This is law.

When I encounter a person that says they do not believe in God, I honor them by not saying a word. I just smile inside for I know what has happened.

In all the years of my dealing with people, I have never met anyone who truly does not believe in God. Some people say, "I do not believe in God" but they do.

How can I be so sure? I test their Crown chakra. If it is open, they believe in God, even if their conscious statements are contradictory.

In the course of our polarized life on Earth, everyone must deny God at least once in their life. There is no other way to have a real, true value of God unless you live without God, even for only a short time. This is the expression of the balance and counter-balance of life.

Because there are very few absolutes, you may one day meet someone who is truly without God in their life. I hope it never happens but, if it does, this is how you may serve them.

Everyone is entitled to a healing. If you deny a healing, you are in judgment of that person. However, you are only to heal them with what is called auric energy. This is the energy that comprises the aura, or energy field, around the physical body.

In this application of healing, you do not actually touch the person. You move your hands slowly down the length of their body, from the top of their head to their feet. Your hands are positioned at least six inches away from the body and follow the outline of the physical structure. This will refresh the energy around the body and will energize the person. The final effect will be similar to vacuuming your carpet to remove dust that has accumulated over a period of time.

5. *"If you consciously take on another person's illness in order to heal them, it becomes a judgment of worthiness in relation to your soul. At that moment, you will have created a heavy karmic experience for yourself."*

Your prime responsibility is always for your own well-being. If you feel that you must take on another's disease in order to heal them, you are expressing a lack of confidence in your abilities as a healer.

This action will weaken you, open your immune system to disease, and might shorten your span of a healthy life. No one can continually absorb diseased energies into their own system and be guaranteed that they will be transmuted into clean energy. There will always be the threat of a remaining residue of their disease, which can make you susceptible to many types of physical disorders.

6. *"It is not in order to apply healing to anyone while they are under the influence of alcohol or mind altering substances."*

Healing is a process that involves the healer and the receiver. Both must be in total control of their mental faculties during the process of the healing. If the healer in not in control of their mind, they cannot effectively disseminate energy to a specific designated area of the body. If the receiver is not in control, they will have great difficulty in accepting the energy and allowing the expression of disease to be healed.

Sometimes this can place the healer in an uncomfortable position. What if someone comes to you for a healing and has had several drinks during lunch? This might create an embarrassing situation for you but you cannot compromise your truth. You must explain the rules and ask them to return at a later date without any alcohol in their system. Most people will understand and be grateful for your remaining in integrity.

If you compromise yourself and feel sorry for them, you are out of order. When in service, you can never compromise the purity of the conditions of healing. If you do it once, the temptation will arise again and again.

7. *"In the expression of the healing energies of God, there must be a balance created called, The Exchange of Completion, to insure completion of the process."*

This healing law exists to insure that the process of healing is completed and balanced to create full effectiveness of the procedure. There must always be what is called an equal exchange of energies. This fulfills the balance of energies between the healer and the receiver. If this does not take place, the energies of the healing could disperse and lose effectiveness.

The exchange does not necessarily need to be money. It can also be in the form of services or goods, but there must be an exchange. I cannot stress the importance of this too strongly.

Example: A psychiatrist called me and told me that his wife had been possessed by spirit for many years. He had taken her to a spiritual healer who proceeded to charge him $5,000.00 for a healing to remove the entity. When he told me this, I knew that the healing did not work. I told him that the exchange was out of order and not in balance.

He asked if he could bring his wife to see me and what would I charge him. I replied, "Whatever you feel is a fair exchange." He did not believe me and tried to make me quote him a definite cost. I finally told him that since there was no price for health, maybe $100,000,000 would not be too much. There was silence on the phone, and I repeated my original offer, "Whatever you feel is a fair exchange."

He brought his wife to see me and I was able to assist in making her well again. When we were finished, he asked me what he owed me. I smiled and repeated my terms once again. He said, "Is $100.00 satisfactory?" I told him that if he felt the $100.00 created a balance of exchange, it was fine. He was satisfied and all was completed.

It does not matter what the final exchange consists of. Someone can share $5.00 with you and it means more to them then another person giving you $500.00. Everything is relative to the value of the exchange in the mind of the client. The important element here is to seal the healing with a balance of exchange.

The laws of healing have been created to protect the free-will of mankind. This can never be violated or infringed upon by another person. By protecting the free-will, the order of the progression of universal karmic evolution of Earth is maintained.

Many healers feel that they are just a vehicle for the healing energies and are not responsible for any effects and results that the healing may produce.

I cannot totally agree with these words. The healer does not control the result of the healing but is responsible for what they do with the energy. The healer is an integral part of the healing process, especially if they have knowledge of the laws of spiritual healing.

It is part of the healer's responsibility to be rested, free of alcohol or chemical substances, and to have pre-set the proper conditions for the healing process to commence.

It is very difficult to fully adhere to the healing laws at all times. Temptation and ego are always a threat for the healer. Take the time to fully understand your role and responsibilities in the healing process.

The concept of life on Earth encompasses the free-will of everyone. There will always be times when you will have to accept an illness in someone, if that is their choice and what they elect to have at any given time.

Absentee Healing

By definition, absentee healing involves the mental projection of energy to a person, place or thing under the following circumstances.

1. To a person not in your physical presence.
2. To a different geographical location.
3. To a situation occurring at the present time in another location.
4. To a person in your presence, who has not requested any form of healing.
5. To animals, plants, or any of God's creations that do not possess human souls.

Before we present you with specific examples, there are certain spiritual aspects of philosophy that are necessary for you to consider as part of this process.
Your current lifetime has been chosen by your soul. It is here to become involved, through you, in experiences that will result in its eternal search for evolution. For this reason, it is important for you to understand that it is the soul's incarnation, or lifetime, not yours! You are the conscious, expressive personality of your soul. You are the one whose actions and decisions constitute the result of your soul's evolution for this lifetime.

The soul is the subtle, unconscious factor that gently inspires and guides you, the person, into the involvement that will ultimately add the proper reactive energy to its existence. With this as the foundation, we can begin to see that disease, or well-being, must always be in accordance with the Divine Order and purposes of the soul.

If someone not in your presence, no matter where they are, requests an absentee healing, without exception always send the healing energy to the person's soul, not the personality. There are no exceptions to this law. By doing this you are never imposing the healing energies on a person. You are making them available to the soul to use at the soul's discretion.

The soul must be the determining factor. The soul must decide if the disease is still needed or not. The soul must be allowed to follow its plan of action for the personality to experience, and to draw its conclusions from that experience.

If you send healing energies to the personality of someone who is ill, you are making judgments. It is like saying, "I have decided that you need to be well."

That is totally out of order. You cannot heal anyone based on your decisions. You are here to serve people through their souls, nothing more. As you begin to transmit absentee healings, it will help you to know what the person looks like or to have a picture of them in your possession as a means of association.

The Recommended Procedure is as Follows:

Lie down and place yourself in a very light state of relaxation. Tell yourself that you are only relaxing your physical body and emotional body. As quietness begins to be with you, you will become aware of a greater ability to concentrate on and with, your mind. When you have become aware of your mind, try to create a thought image of the person and sense their energies with your mind. If you are unable to picture the person, make the thought statement: "I am here to be aware of the energies of. _____ (name)."

Lie quietly for several minutes and allow yourself to slowly activate your mind. Open your Crown chakra as you breathe. Draw energy into the chakra and send it to your brain from this spiritual center.

You will begin to feel as if you are gathering a mass of energy together in your mind as a ball of power. As the buildup of energy grows, make the following statement: "I project the healing energies of God to the soul of _____ (name). The soul may use these energies at its own discretion and for the Divine Order of its purpose at this time. If it is within the order of the soul, then allow _____ (name) to receive God's energies so that a healing may be effected."

Concentrate quietly and project the ball of Light from your mind to the person's soul. Feel it leaving your mind as you exhale and try to sense a blending between the energies and the other person's soul. When you sense that the connection has been completed, relax and gently open your eyes. The process of absentee healing has been completed. I have described a general method of absentee healing application. There will be times when exceptions will occur but, for the most part, this will fulfill your absentee healing expressions.

Case Studies

1.   Years ago, a friend told me that her sister had become inflicted with an inoperable brain tumor. She was given only three months to live. My friend asked me to send a healing to her sister. I told her that her sister would have to ask me herself, according to the Laws of Healing.

The following week I received a beautiful letter from her sister. She told me that she did not know about my work but she believed in the power of God to heal. She stated that she still had much to accomplish in her life and asked me for the healing.

I followed the absentee healing procedure and several weeks later received a call from the woman. She was overjoyed. Her physician told her that there had been a total remission of the tumor and her life was no longer in danger.

Five years later, my friend informed me that her sister had developed another brain tumor and asked me to send her another healing. I told her that I was sorry, but I could not do that. Her sister knew the proper procedure and had not contacted me in relation to the new problem.

I told her that this time, her sister was ready to leave life here and return to God. If not, she would have contacted me to assist her once again. This was one of those rare situations where the healer must be detached, objective, and never compromise the healing laws. It becomes easier if we remember that life here on Earth is the life of the eternal soul. The personality is just a temporary expression that serves the soul and its evolution.

2. A couple called me to request a healing for their child who was afflicted with cerebral palsy. I responded to their request, as the child was too young to be responsible for this type of decision. (See the Laws of Healing.) I sent energy to the child's soul every day for one week. On the final day, I intuitively received a message informing me that the child's soul blessed me for the energy. It also informed me that it had purposely chosen that physical condition of disability for that lifetime and would remain in that condition for the duration of the life.

This raises an important issue. Did I have the right to say; "I am still going to try to heal this child", or was I to honor the soul and discontinue the healings? I had to honor the soul. I did not like that decision, as I wanted to help the child, but I had to heed Universal Law.

I knew inside, that under those circumstances, the healing would not have helped the child to recover. I sat down with the parents and gently explained the circumstances to them. I told them that their child had an important purpose on Earth. She was here to help people to love her, not to pity her; to accept her as she was; to look beyond her disabled body and sense and love her soul.

The parents understood and from that moment the whole family began to learn from the child. They began to see what love truly was and stopped all the self-judgments, the blame, and expressions of self-guilt and pity.

The finite expression of absentee healing takes many years to achieve. It requires being in a state of absolute knowing that your mind expression creates the results. Until you have reached this degree of expression, you hope. You try to believe it can work. When you KNOW, everything changes and you are one with your mind and the Universal Light.

For example: A person calls you and requests an absentee healing. If you are in a state of knowing, all you need to do is to acknowledge their need, and say to them, "Consider it done." This is finite healing. You know that healing is a polarity of a request and a response. And under those conditions of the action-request and the reaction acknowledgement, the balance is restored.

To reach the state of knowing takes many years of practice and the confirmation of the results of your healings. The greater the degree of your success, the less doubt arises in your mind. The time comes when you know that a request and a response heal.

Absentee healing is available to everyone. If you see a disabled person walking on the street, have a thought to send healing energy to their soul.

If you pass an accident, send absentee healing to the souls of all involved in the accident. If someone is in the hospital, send the soul energy before any surgery and after surgery to quicken the healing process.

Perhaps the ultimate example I can relate to you was the time I received a call from a woman whose child was stricken with a severe nerve inflammation. She asked me to send healing to her child. I told her that at six o'clock every evening to lay down with her child, and I would be there with healing energy for seven days.

By the next day, I had forgotten about the phone call. A week later she called me to thank me for the healings. She said that she felt my presence and a surge of energy every day at the appointed time.

I had done nothing consciously after our first conversation but I knew that my statement of intention would produce the results that took place.

Establish the condition of knowing as your ultimate goal in establishing the control of your mind and your emotional reactive system. As long as you send the healing energy to the soul, everything will always remain in Order. The longer you are involved in healing, the more you begin to realize that your service has a dual expression. You serve the person and their soul.

It is difficult, at times, not to question the purpose or reason for certain expressions of disease. We want everyone to be well. When you know that, in the ultimate truth, life is the soul's adventure, you will have the strength not to question how and why, but always to remain in the Universal Law.

Healing Energy Aberrations

The discussion of unusual energy conditions can be difficult to understand and to accept as expressions of life. There are so many conditions that we have no conscious explanations for and yet when we are dealing with energy, reactive expressions begin to reveal knowledge about areas and causes of dysfunctional behavior.

If you can assist people in the understanding and recovery from expressions of energy aberrations, it will be a very rewarding service that you, as a healer, can provide.

As we begin this discussion, it must be noted that the majority of people are not affected by these conditions of life. It is an accepted fact that everyone has been exposed to some form of dysfunctional behavior during the course of their life. We all have expressed some of the symptoms that I am going to describe. Normally, we deal with them during the time of our daily functions.

The conditions we are dealing with manifest after prolonged exposure and repetition have created the acceptance and expression of these energy patterns. There are three distinct categories of aberrations that are consciously created and expressed. Each of them can result in causing major problems in a person's life if they are not recognized and dealt with.

1. Thought form energy invasion.
2. Self-obsession.
3. Soul invasion or possession.

Thought Form Energy Invasion

Every time someone expresses a negative thought to you or about you, they have generated an energy cell of negativity. This is projected to you and enters your aura or outer energy field. If you allow yourself to be affected by the words or actions, the negative energy cells can enter your physical body. In effect, you are opening a door and inviting this negativity inside.

It then searches for a home in either your solar plexus or second chakra. These centers are the main repositories for emotional and worthiness energies in the body.

When the negativity that someone imposes on a person is not resolved, it will remain alive inside of the body. Every time a similar incident occurs, the negative energy gains more power. In time, this mass of energy can develop into what we call a black snake inside of the body. The resulting effects of this expression can produce unhealthy attitudes in life and relationships.

If someone has experienced a negative childhood environment without any expression of love or worthiness or if they were the victim of repeated dysfunctional behavior, these energies could be created, enter the body, and remain there, altering the normal expression for life.

As the person would mature and encounter life situations without success or self-appreciation, the energies of negativity could become stronger and stronger. In time, there would be a serious effect on the conscious behavior patterns.

The most common symptoms of this condition are:
- Ulcers
- Colitis
- Constant unworthiness.
- Resenting success of others.
- Sudden outbursts of anger or depression.
- Nightmares.
- Gradual change in behavior patterns.

If the negative energy is present in the aura around the body, a sweeping of the aura will remove the energy before it has a chance to enter the body structure. We all pick up some negativity during the day's association with people. You can sense this easily, if your body feels a little tired or heavy at the end of your day's activities.

To sweep or cleanse the aura of negative energy, we suggest the following procedure. Have the person stand quietly with their eyes closed.

This enables them to feel the energy as you conduct the healing. Create the mind-thought that your hands will serve as a magnet to pull the negative energy out of their aura to the surface of your hands. By setting the limitation of pulling the negative energy to the surface of your hands, you prevent the energy from entering your body. It is important to establish this limitation for all applications of healing. When the healing has been completed, rub your hands together, have the thought that you are transforming the energy into Light, and all will be in order.

Hold both of your hands above their head, approximately six inches or fifteen centimeters away from the body. Slowly move your hands down following the shape of the body. Your mind-condition is that your hands are magnets pulling down all negative energy. When you reach the floor, rub your hands together, and transmute the energy.

Move to the side of the body and repeat the procedure. Sweep the aura from four positions as you move around the body. In each position, transmute the energy you have pulled down the aura. The process will only take five minutes to complete. When you have finished, the person should feel lighter and more flexible with their normal body movements.

In order to determine the location of negative thought-energy inside the body, the following procedure is recommended.
Have the person lie down. Place your hands with the palms down, about three inches or seven centimeters above the body. Your mind-intention is to feel the body energy on the surface of your hands, without allowing this energy to enter your body.

Move your hands very slowly over the entire length of the body. You are trying to locate areas where heat is coming out of the body to the surface of your hands. This examination will help you by indicating the general location of the negative thought-form energy.

The suggested procedure to remove the heat energy is as follows:

a) Have the person begin to breathe deeply through their mouth. As they exhale, let them feel as if they are pushing the energy down the length of their body.

b) Place both of your hands on top of the mass of heat energy and begin to massage the area as if you were kneading dough. Slowly gather it together into a small ball. With a thought, make your hands magnets and pull it directly out of the body to the surface of your hands.

c) Hold the ball of energy in your hands, have the thought to transmute it into Light energy and release it.

d) Place your hands over that area of the body and begin to send healing energy into that area. This will replace what you have removed with positive energy inside of the energy center.

e) Have the person breathe thought-love energy into that area to establish a new pattern of worthiness and self-acceptance.

It has been my experience that sometimes the body will feel tender in that area for several days. Some people will go through a cleansing process that could be similar to the flu. This is a form of detoxification that will re-balance the associated area of the body.

Self–Obsession

This form of energy aberration is often mistaken for a condition of soul invasion or possession. The resulting symptoms are often similar, making it difficult to determine the existing condition.

Everyone talks to themselves periodically. We fantasize, daydream, and create magical illusions that are so wonderful for us. Sometimes we even create a little voice inside our head to help us make decisions. There is nothing wrong with this. In fact, it helps us in our creativity and for future motivation to reach new goals. Many times talking to ourselves can help us to make the proper choices for our actions in life.

There are times when we generate negative thoughts in our mind. If we choose, we can use those negative thoughts to avoid situations and actions. By doing this, we are able to eliminate the chance of making a mistake or experiencing failure.

We have the voice tell us exactly what we want to hear. That is perfect. Now we don't have to blame ourselves for taking an improper action. Now the problem can begin to take root inside of our thought patterns. If a person begins to depend and rely on this inner voice, they could begin to create what is the true expression of an alter ego. It could result in the beginning of a dual expression or split personality.

There is a real danger present. The inner voice could become stronger and stronger. This could result in the person giving up their freedom of choice and the responsibility of making their own decisions. The inner voice could gain strength and become the dominating factor of conscious expression. If this happens, a true split personality could result.

What is most important to remember is that, in these instances, there are no outside energy influences present causing a problem or improper condition. Everything has been activated from within the person and is self-contained.

Because of these circumstances, there is nothing to be removed. There is no need for any type of energy healing to correct this situation. Everything belongs to the conscious personality.

The focus of the healing process is in the counseling. The aim is to restore the balance inside and enable the conscious mind to regain control of all thoughts and decision-making.

If a person tells you that they hear voices inside of their head, ask them to describe the thoughts and words. The majority of the time you will be able to determine from the conversation whether this is a self-induced condition. If it is self-induced, I recommend the following course of action for treating the self-obsession.

1. You must carefully explain that they have created this condition themselves. Without their permission, the condition could never exist. Many people will assume that it is an outside force doing this to them.

2. The person must be willing to again assume control of their life and be responsible for the result of all actions. If they are not willing to agree to do this, there is nothing further you can do for them.

3. If all conditions are understood and acceptable to the person, explain the following procedure to them.

Every time they hear a voice, they are to interrupt and say, "Go away. I don't need you any longer." The first day, they may have to repeat those words a hundred times. It does not matter. They need to have a lot of strength and desire if they wish to become whole again.

The second day they will hear the voice less often. Every following day, they will hear it less and less. By the end of thirty days, the voice should have totally disappeared. Each time they deny the need for the voice; they are re-programming their sub-conscious and reclaiming the control of their conscious mind.

This will not be easy for them to do. It takes a great deal of courage and strength to negate a sub-conscious program but it is the most successful method I have discovered to re-establish the conscious control over a life.

Tell them that you are available as a support system for them, if they need it during the course of this process. This is important for them to know. Creating change generates a lot of insecurity and having encouragement from you will help them to continue the process to completion.

Soul Invasion or Possession

By definition, this condition involves the entry into a person's physical body by a soul whose purpose is to gain control of the conscious mind and the lifetime. Another way to state this would be, "As long as a person's aura remains sealed, they cannot be invaded by outside energies or other souls." If you remain in conscious control of yourself, nothing can invade or enter your body!

Because of these conditions, most possessions occur in places where people gather and socialize and have too much to drink — mainly bars and social clubs. The susceptibility is also present in places where narcotics and chemical substances, such as drugs are used to excess.

Souls that are looking for a body to invade will usually gather at those places where people have lost the total control of conscious mind and will. This condition weakens the power or seal of the aura, enabling the outside, alien soul to invade the physical body.

People that lead their lives in lies or continual sacrifice of their own value and worth may also weaken the seal of their auric field of protective energy.

There are many conditions that can lead to a possession taking place.

The most common situations are:

1. A person dies suddenly and unexpectedly. This can cause a shock to their soul. It no longer has a body or a life. The pattern of emotions is still active and the soul can become concerned for its family, business, house, etc. It may panic and look for another body to take over and continue its planned lifetime.

2. A soul is angry and wants to satisfy that anger. It will enter a body and express strong destructive behavior. Its purpose is to force out the original soul and use the body and life for its own purposes.

3. Many possessions take place because a soul just likes it here on Earth and does not want to leave the emotional plane of life. It will enter a body, find a safe place, and just peacefully hide itself. It does not interfere with the person's life and remains unnoticed for as long as it wishes to remain inside.

4. The majority of possessions are created by souls that want to continue life here on Earth. They do not intend to cause any harm but, as time passes, they gradually assume more and more control of the conscious mind and life.

Any form of possession is in violation of the Universal Law of Conscious Life. The soul will generate a very severe karmic reaction in its energy structure. This action is a total judgment, as it invades and damages the life path of an incarnated soul.

How do you know if someone is possessed? Most people, who come to me for this problem, have the feeling that someone is inside of them. It is very rare that a person will be totally oblivious to this type of condition.

There are certain signals that indicate a soul invasion.
- Seeing anger or hate coming from their eyes to you.
- Avoiding eye contact with you.
- Consistent manic-depressive behavior.
- Sudden outbursts of anger without warning.
- Consistent thoughts of suicide.
- Constant state of anxiety.
- Speaking aggressively most of the time without justification.
- The constant presence of negative, destructive thoughts.
- The presence of negative voices that increase in power with time.

Now that you have read this list, do not begin to wonder if the list applies to you or associate it with people that you know. Please remember that everyone participates in some of these expressions during the course of life and, certainly, on a daily basis to some degree.

The resulting condition of soul invasion or possession affects a very small percentage of people. This is not a common condition that one encounters on a regular basis.

If I deal with four or five people a year with this condition, it is about the average. And I see hundreds of people a year.
If you have concluded that a person has been invaded by another soul, you must proceed with a counseling session. There are several critical questions for you to ask. This will greatly assist you in determining the state of mind of the afflicted person, as well as their outlook for their future life.

1. Are they willing to re-assume control over their life?
2. Do they really want to be free of the possessing soul?
3. Do they enjoy the company of the other soul?
4. Can they be responsible for their actions and decisions in daily life?

I want to make you aware that some people enjoy this presence inside of themselves. They are lonely and finally have something to relate to. If they are happy the way they are, leave them alone. They are entitled to have their choice of life. If you remove a possession in them, it will just return when they leave your presence.

The main key question is: Are you ready to be free and become responsible for your total expression of life? If they hesitate before answering, tell them to go home and decide what they really want. You would be surprised at the number of people who will keep the possession. For them, a void in their lives has been filled. They do not desire to live and express their lives in the reality of the structure of society.

If the person has stated a true need to be free and wants you to be the instrument for their freedom, I suggest the following procedure.

When a person has been invaded by another soul, it is always of benefit to locate the areas of the body that contain a concentration of this soul's energies. The more knowledge you have beforehand, the more confident you will be during the clearing process.

Possessing souls usually concentrate themselves in the expressive centers of the body. In this way, they maintain a connection to the person's emotional system. This connection feeds them and increases their power.

I am going to list the centers that are the concentration places of possessive energies.

- The genital area or base chakra.
- The second chakra. The emotional reactive center.
- Breast tissue.
- Throat.
- Solar Plexus.
- Upper thighs.
- Knees.

Have the person lie down on their back and devote several minutes to place them in a relaxed and peaceful state of body and mind.

It is now time to establish the conditions for the clearing. Stand next to them, close your eyes, and repeat these conditions silently with the power of your mind.

1. Ask God to send you sufficient power to remove the soul that has invaded the body.

2. Ask God to shield and protect you with Light to keep your aura protected and sealed against invasion.

3. Ask your spiritual masters to be with you to assist you during the process.

4. Ask for the spiritual masters of the possessing soul to be present and to take the soul to the level of heaven where it will be healed.

(This is very important. This will insure that the soul you remove will not remain in Earth's vibrations and attach itself to anyone else.)

I must make you aware that you cannot be in judgment of the possessing soul. You cannot express any anger or use any expression of negativity in the removal of the soul. The process must be conducted with feelings of love and an attitude of Grace.

I cannot stress the importance of taking this attitude. This process requires absolute detachment, and personal dedication to the well-being of both souls. If not, you will not be able to complete the process. If you display any fear, you empower the possessing soul and it will successfully resist your attempts to remove it from the person.

Clearing Process

Place your hands on the person's ankles and very slowly and gently, move them up the entire length of the body to the top of the head. The purpose for this is to send energy into the body to prepare for the removal of the invading soul.

Return your hands to the ankles and hold them firmly with a slight pressure. Focus your mind; make your hands like magnets and set the condition that you are going to pull the invading soul's energy all the way up the body and out of the top of the head. Hold the ankles firmly and very slowly begin to pull the energies higher and higher up the length of the body.

When you reach the shoulders, mentally hold the energy in place. Put your hands on the person's fingers and pull the energies up the length of their arms to join the rest of the energy there.

Now it is time to establish the final conditions for the clearing. As you continue to pull the energies up towards the head, say the following, "When I count to three, the soul will be out of the body and will be taken by its spiritual masters out of Earth's vibrations to the level of Heaven where it can be healed."
As you reach the top of the head, begin to count. When you say three, cup the soul in your hands and, symbolically, release it to the masters that are there to serve you and assist the soul. The clearing is done.

Immediately touch all of the person's seven chakras with a hand and seal them. Cover the person with a blanket and have them rest quietly for twenty minutes. Hold their hand and speak softly and gently to them. Reassure them and comfort them.

Now that the clearing has been accomplished, the balance of your work is in front of you. The presence of an alien soul has created chaos and weakened the person's entire energy system. It is your responsibility to rebuild the strength of their energy meridian structure.

Place a hand on their Crown chakra and send God's Light into the chakra for two minutes. Repeat this process on the balance of the six other chakras. Return your hand to the Crown chakra and slowly move it down the center of the body connecting all the chakras together. Do this process seven times all together. The purpose for this is to pull the chakras into alignment and create a balanced flow throughout the immune system.

It is recommended that you conduct daily healings for at least one week if possible. This will help to strengthen the complete energy meridian system.

Because of the soul invasion, the body is going to detoxify itself. The person may develop flu symptoms; have a fever, upset stomach, etc. This helps to cleanse and purify the body and complete the inner healing process.

Sometimes a person has been possessed by more than one soul. This is very rare, but it does occur. If a person has been invaded and enjoys the company, they may invite others to come inside. This is a reality!

I had a woman come to see me with thirty-six souls inside of her. She was happy with her friends, knew their names and their likes and dislikes. They filled an empty void in her lonely life. She came to me when she was ready to let them go out of her life and I fulfilled her request.

When you are doing a clearing, it is no more difficult than reading the words I have written describing the process.

It is not like the movies. You will rarely encounter resistance and certainly will not experience violence of any nature. It will be just like any other healing, calm and peaceful.

Since this may be a new experience for you, I advise that before you undertake this type of healing, you observe several that are conducted by someone who is experienced in this area of service. This will enable you to observe the process without creating any inner fear or apprehension about the process and the healing.

I have performed several hundred clearings of this nature over the years. I have never experienced any violence or attempted attack taking place. As long as you establish the proper conditions and protection I have suggested and stay in Grace, you will have no concerns or problems.

Case History

Many years ago, I received a call from a man in California. His wife had been demonstrating split-personality behavior for several years. It has progressed to the point where she had tried to set fire to the house and to kill herself.

Before he called me, he had taken her to a hypnotist. Under hypnosis, the invading soul took control and began to speak out loud to the hypnotist. The man did not know what to do and ended the session. That is when he called me.

He brought his wife to see me. She was very coherent but was in a state of deep confusion. I placed her in a deep relaxed state and asked the invading soul to speak to me. I will relate relevant parts of the conversation to you.

This was a very angry soul, who blamed the woman's soul for her death in a past incarnation. This soul invaded the woman to destroy her as revenge. I told the soul that this was out of Universal Order and that I was going to remove it from the woman's body and send it to a place where it would receive love. The soul asked me, "What is love?"

When I explained love and the procedure I would follow, the soul agreed to leave peacefully. That is exactly what took place.

When the clearing had been completed, the woman sat up, smiled, and said, "I feel like I have just been born." I completed the healing process as I have described and sent them home. I received Christmas cards from them for years. She is well and leads a productive life.

This case was a complicated situation that was in a very advanced stage of expression and yet there was no violence. All was dealt with in peace, love, and Grace.

For me, a true miracle had taken place. How can I describe adequately to you the feeling of humbleness I experienced before the presence of God's Light? Many times, the result of a clearing has literally saved a person from spending extended time in a mental facility, if not for the balance of their life.

As long as you are involved in healing, always remember that you are in the service of souls. There cannot be any judgment or anger, regardless of their behavior. You cannot allow yourself to feel sorry for someone, to feel pity or feel sympathy for them. You are present to serve them. Remain detached and fulfill your commitment to the healing Laws of the Universe.

All healings must be undertaken with this understanding: If you are to properly serve others and walk in the truth of your soul and God, stay in Grace. And you will not have any problems or concerns.

Accidents

An accident is created from the result of any action whose expression is not the normally anticipated or the expected result of that action.

For example: You are taking a walk and twist your ankle on a stone. The action is walking. The twisting of your ankle is the unexpected result that is not the normal result of taking a walk.

You are trying to hang a picture on the wall. As you try to hit a nail in the wall, you hit your finger with the hammer.

Some people would call both situations instant karma. For me, this is an illusion and out of reality. You twist your ankle because you are not watching where you are walking. You hit your finger because you don't know how to hit a nail properly. We need to be real. Everything in life does not have a spiritual, karmic purpose.

If we can accept that the universe is always in total Divine Order, we must believe that, spiritually, there are no accidents. This evolves into the statement that eternal spiritual energy must also be in Divine Order.

From the moment a soul is created by God, it follows a progressive plan of evolution. We could compare this to our educational system, where a child goes to many levels of school. In the different schools, the child acquires knowledge and develops a philosophy of truth and expression. We call this the karmic path of evolution. As a result of these experiences and education, the soul determines its interpretation of the Universal law.

As a result of the many people who have come to me for healings or classes, I firmly believe that, spiritually, you are drawn to people at the proper time, to the places you are to be, and to the circumstances that you are to experience. You are led to the teachers that will contribute to your growth and also to those who are not valid for you. This polarity of experience enables you to experience, discern, and establish your truth. The same formula will apply to your physical, emotional, mental and spiritual well-being.

There are many spiritually oriented people who do not believe in any form of accidents. They believe that every action, either physical or spiritual, has been orchestrated by the Divine Order.

I have always found this difficult to accept in the reality of the laws of free-will and choice that are inherent in our lives. All actions are directly related to the energy of karma, meaning experience. This makes sense. An action always produces a reaction that results in an experience or karma. The words are truly interchangeable. Any experience becomes a lesson when you decide that you have something to learn from the result of that experience. This decision continues the karmic experience of growth in the formulation of truth resulting from that particular action.

Our emotional energy and expression are only temporary tools of expression. They are limited to the Earth plane and the pattern of karmic life of Earth. Beyond the limits of Earth energy — the astral planes — emotional energies do not exist. They are purely tools for Earth karmic experience and evolution. In the outer dimensions are the pure energies of Divine Order, nothing else.

Is it really an accident if:
- You are a drunk driver and hit me?
- You are mentally unbalanced and shoot me?
- You steal from me?
- You fire me to give my job to a relative?

Perhaps we might consider that, in these circumstances, I have become a victim of your free-will expression!

The free-will laws of planet Earth allow for our variable expression and experience. We are affected by the free-will of others when it is imposed on us without our consent or agreement. Free-will is not a part of soul expression. It is limited to the conscious, karmic experiences of life on Earth.

The results of the questions I stated to you, created karmic experiences as a result of the actions. They were not the cause of the actions.

Case Histories

1. In 1974, my friend's thirteen-year-old daughter was crossing the street. An approaching vehicle did not see her. She was struck by the car and killed. The next day her soul came to me and spoke to me. Her soul said that she was not supposed to die at that time. She had not even begun to experience what she had come here to do. The soul said, "It was an accident", and it was returning to Earth, in a new life, in only several months.

2. During a meditation, I was interrupted by the presence of an unfamiliar soul. His energies came into my thoughts. He was frightened and confused. He said that he had heard a noise and all of a sudden, he was without his body. He lived in Ireland and had been shot during an insurrection. The soul was in total shock. He said that he still had planned much to do during the lifetime and was supposed to live for many years to come.

Over the course of time, I have had numerous experiences of this same set of circumstances. There have been too many souls crying, "It was an accident. It was not supposed to happen to me." The number of similar cases has convinced me of the reality of free-will accidents of a physical nature.

A child, who is too young to be responsible for its actions and for its life, is always affected by the free-will of its parents. A person, who is sickly or advanced in age, is subject to the free-will of the people responsible for their care and health.

As time passes and a spiritual awareness begins to come into our lives, we experience a totally different pattern of life. Spiritually, I believe that accidents do not exist. When we understand that the true, eternal existence of the soul is in spirit energy, then we can accept the Divine Order of all things and energies in the Universe. Under this premise, spiritual accidents cannot occur, as the Universe must always be in total order.

I believe that we meet the people we are supposed to meet. The only variable is time.

I believe that we go to the places necessary for us to experience.

I believe we take the actions to be in the right profession for our growth and learning.

The key is always time. For your soul, for the Universe and eternal existence there is no time as we know it. All energy just is. No past, no future that is an unknown. Everything exists now, as energy is eternally constant. It does not fade or disappear.

When the concept of eternal time becomes acceptable to you, the pressures of life lessen and you begin to understand that it does not matter when things happen, just that they do take place during your life.

It has been important to briefly discuss accidents with you, to enable you to more accurately validate the causes of disease or injury to the body. This will assist you in selecting the proper methods of energy application during your counseling and healing sessions.

Karmic Disabilities

Before I begin to discuss this area of life, I feel it is important to make the following statement.

I cannot prove the thoughts and feelings that I am going to express to you in this chapter of the book. What I am going to say to you is my philosophy that I have accepted after many years of close involvement with people and the situations that have expressed themselves in their lives. Each person must adopt their own truth and comfort level in these unusual circumstances of life. If I accomplish nothing more than to initiate a thinking process for you, I will have achieved my purpose for this writing.

A karmic disability is any physical, mental or emotional condition that exists from birth, and will be present for the duration of the lifetime.

It is necessary to present the manner in which I relate to the word karma. Many people interpret this word to relate to a form of punishment for a past action in either the current life or a past lifetime. If we were to accept this, it would also mean that we have accepted the reality of judgment and punishment in our lives.

In my philosophy, the word karma means experience, nothing more. It has no connotation of being either positive or negative, just an experience of life. When an action has been taken, we have created an experience, or karma, from which we are supposed to decide, "What is my truth?" in that situation.

I believe this is the total purpose for karma. The moment we decided what is our truth and realized what we learned from the experience, the energies of that karmic experience have been completed. We are free to move on to the next experience of life.

When a soul is created by God, it begins a process of learning and evolution. We could say that the soul goes to school to learn the laws of God and of the Universe.

Certain levels of this school involve the necessity of a physical form of expression called incarnation. The planet Earth is a level where the classes are involved with the expressions of free-will and emotional interchange with other people.

Philosophy tells us that the soul chooses each incarnation and, basically, designs it so that it will be exposed to all the conditions and circumstances that it has chosen to experience. The results of all the circumstances are actions that create the karmic experience of growth for the soul.

With this as a foundation, I say to you that when a child has been born with a physical, mental or emotional disability, most of the time the soul has chosen these conditions of life. It is what we call a minority incarnation.

I acknowledge that these words are very difficult to accept and to understand. We must ask ourselves why would a soul voluntarily choose a disabled lifetime. I will try to explain this to you in the best manner possible.

It does not matter what expression the disability takes — mental retardation, Down's syndrome, autistic expression, or physical deformity. The purpose for the soul is all the same. The karmic experience is all the same. A lifetime of service to all people it encounters, has a relationship with, and is part of the immediate family.

For these reasons, we are dealing with a strong, developed soul. To take an incarnation where the soul will have great difficulty reaching the conscious mind requires great strength and power within the soul itself.

The ultimate purpose for this category of incarnation is to teach people that the true existence of life is the soul or what is really inside the physical shell of life. It is not who or what people seem to be.

This is a most powerful lesson for people. We look at those people and feel pity, shame, we avoid them, and we are even repulsed by them at times. We judge them according to our own values and egos, without bothering to consider who they are and why they are in that condition of life.

All types and expressions of this category of disease are for the duration of the lifetime. Healing that can result in a possible cure cannot be effective. The healing process that should be utilized is to mentally project energy to the person's soul.

There are many reasons for this.

1. It serves as recognition of the existence of the soul, as well as its presence inside a physical body.
2. You are acknowledging that it is the soul's lifetime, not just the personality.
3. You make energy available to the soul to assist it in expressing and completing the purposes for which it elected that incarnation.
4. The soul begins to sense that it is not alone; that someone else truly cares and expresses love.

When you meet someone who is in a minority lifetime, do not judge them by whom they seem to be. Look inside. Feel who they are and allow yourself to relate to that center core of life, their soul.

Many years ago during a healing seminar, one of the students asked me if a friend could bring her six-year-old child there for a healing. The child was afflicted with cerebral palsy. The woman brought in her child the next day. Everyone was sitting as part of a circle and as the child entered the room and came up to me, I watched the people's reactions of sadness and pity.

The little girl came over to me and I sat her on my lap. I asked her if she knew why she was there. She smiled, turned to face the students in the circle, and raised her hands with her palms out towards everyone in the circle. She then proceeded to send healing energies to everyone. When she was finished, she kissed me, and walked out of the room.

Everyone had tears in their eyes, tears of their shame for judging and pitying the young child. They had all created a situation called, instant karma and learned from it instantly. They learned that the true beauty of a person is not their physical appearance but comes from the inner house of the soul.

I cannot present a better example of a service soul that is confined inside of a body with a karmic disability.

If someone should bring a disabled child to you for a healing, the greatest service you can perform is to teach them how to communicate with their child's soul. This process is identical to the process for absentee healing. By taking this action, the parents may be able to become aware of the true purpose for their child's disability and what their role could be in assisting the child's soul in completing and fulfilling its purpose for life on Earth.

You will not be able to effect a healing cure but you might be able to stop the progression of the condition and stabilize the physical body in its current stage of disability.

Imagine what a gift it would be, if the soul became aware of a parents thought energies of communication and was no longer totally isolated from the outside world, unable to achieve any progress in evolution and experience.

Physical and Spiritual Healing

When a soul becomes embodied in the physical creation of God, it has become an integral part of the expression of physical perfection. The physical structure must be considered as the temple that houses the soul. Therefore, the body and all of its components and organs are to be honored and valued in a similar manner.

There are no parts of the body that are dirty, unworthy, too small or too large, or unmentionable. Everything, and I mean everything, is to be an honored part of the complete expression of your soul!

This sounds so wonderful. I wish it were being practiced and lived by everyone. If it were, there would not be any disease, none whatsoever, as our bodies are created and structured to live for many hundreds of Earth years. The proof of this is recorded in every scripture that has been written.

We could compare ourselves to a computer. Our soul is the programmer, our conscious mind is the viewing screen, and our subconscious serves as the memory disc. Our soul programs expressive patterns of its planned behavior for the conscious mind to view on the screen. If they are acceptable, they are recorded and stored as positive actions and responses for further reference in the subconscious mind.

What happens when the soul projects a thought to the conscious screen such as, "You are a worthy person", and the conscious personality says, "I am not worthy"? This is what becomes the result.

The conscious expression of unworthiness becomes recorded in the subconsciousness as the truth input from mind. If this data is not compatible with the current state of expression, an adjustment is initiated to reprogram the truth between mind and body. This is how we can find ourselves being in a state of disease. The reprogramming could make the mind's statement of unworthiness the truth. For this reason, I make the statement to you that I will repeat many times: OUR EXISTENCE IS MIND!

If we can begin to accept these words, we need to consider that every imperfection in us has been triggered by a mental statement to ourselves.

For example:
- I am not worthy of love.
- I hate my body.
- I don't deserve happiness.
- I don't love myself.
- Parts of my body are shameful.
- I know that I am here to be punished.
- I can't be or do ...
- I sacrifice myself at my own expense so others will like me.
- I judge myself after my actions have been completed.

The Law tells us: As you judge yourself or others, so it shall return to you tenfold increased, until the need for judgment has been transmuted into Grace. Only then can you be healed and cured. If, at some point in your life, you can live in the truth of these words, you will walk with freedom, peace, and joy in your life. Every other technique you try is only an illusion that eventually brings you pain and disease.

Physical Healing

Physical healing involves treating a condition that has been created by the results of an action by yourself or another that has an effect on your physical body.

Examples of this are as follows:
- Tired muscles after exercise.
- A strained or pulled muscle.
- A broken bone.
- Fatigue.
- Low energy resulting from work or play.

There are many more but let me classify them all as conditions that do not have any spiritual implications.

All other expressions of imperfection or abnormal conditions can be considered to have a spiritual causative factor involved.

Whenever we are dealing with a pure, physical healing, we must remember that our role is to supply energy to the body for that specific condition. All comes from the mind, so we have the thought, "I am sending energy to heal fatigue, tired muscles, etc." It is necessary to always state the purpose and desired result as the goal for sending the energy for healing. If we leave it to God, nothing happens. We, not God, are totally responsible for the healing process.

Every organ in our body has a polarized field of energy that surrounds it and maintains its health. An action that causes a physical condition can affect that field of energy. It makes sense that, if you work muscles too hard, the energy field will become tired and weak.

In all cases of physical healing, your role is only to supply energy to re-strengthen the weakened energy field in the area of the body that is affected. When you concentrate your mind on one specific purpose, our results will be much more effective. There are times when all that will be needed will be a healing of the auric field around the body. This will stimulate all energy that enters the body and begin to refresh the individual, returning them to a normal condition.

If a person comes to you with a broken bone, do not wiggle your finger and expect to fix the bone with energy. The truth is that you are not E.T. and would be wasting your time. Send the person to the doctor to set the bone. You aid in the healing process after the bone has been re-set.

When a bone is broken, a tear develops in the energy field surrounding that area. When the bone is set, the tear is still present and, if left alone, will take a long period of time to repair itself. If you were to project energy mentally to that area, it might assist in the mending process by supplying the bones with a supply of healthy energy. This could shorten the length of time required for the healing to be complete.

Example

A woman came to see me with a massive malignancy in her left breast. She did not wish to have surgery, and wanted to come to me for healings. I told her that she was endangering her life, and to see a physician. To make a complicated story short, she had a radical mastectomy and the doctor told her that it would be at least six months before she would be able to raise her left arm to any substantial degree.

After the surgery, she came to me for healings. I supplied energy to the tears in the energy fields that were affected by the surgery. In a period of ten days, she was able to reach over her head with the arm, and was driving her car without any problems.

I wish to remind you that my total role was to make the energy available to that specific area of her body. Her strong will and desire were the catalysts that enabled her body to utilize the energy to the fullest extent. This resulted in the shortened period of time for recovery.

Example

A professional weight lifter came to see me and asked if I could suggest a method by which his body could recover its energy after a workout in a shorter period of time.

I moved my hand over his aura, about five inches, or twelve centimeters away from his body and energized his aura. This took about five minutes and his energy level was restored to full power. I told him that in the future, after a workout, to lie down, close his eyes, and mentally begin to draw energy into his body through his Crown chakra. Then, as soon as he would begin to feel the energy that he moves it mentally down throughout his body for a period of five minutes.

He utilized this technique at the conclusion of every workout with very satisfactory results every time. The power of the MIND!

The time will come in the future, when we will be able to heal broken bones and regenerate diseased tissue with energy healing.

At this time, we do not have the level of consciousness to absolutely believe this! That is the magic catalyst — total, absolute belief on the part of the healer and the afflicted person. As long as we remain in our emotional patterns of unworthiness and our doom-orientation, the ability to effect total health will continue to be beyond our conscious capabilities.

Many years ago, a man and his wife showed up at my door. They have driven several thousand miles to come and see me. They were simple farmers with gentle and simple thoughts about life and God. He was afflicted with a rare disease, elephantitis. The bones in his arms and face had begun to grow and deformed his appearance.

He said that he KNEW I could heal his condition. That was why he traveled so far to see me. He was not in any anger or judgment of God but felt that he was an instrument and example for many people. I gave him a very intensive healing and prayed that, if it was the will of God and his soul, he recover from this expression of disease.

When the healing was finished, he thanked me. He left, and his parting words were, "God sent me to you and I know I will be well." I had tears in my eyes but I believed that a miracle had taken place.

Six months later, he showed up at my door again. I did not recognize him or know who he was. He smiled at me and said, "Look at me, I am totally recovered." I could not believe my eyes but his structure had become normal. I wept. We wept together and thanked God for Its blessings.

There have been many other instances of total healing. There are also many more times when nothing happened. The healer cannot question or ask why one person recovers and others do not. That is not within the Law. The outcome of service is the decision of the soul and God. We are not to question, doubt our abilities, or lose faith. We are here to serve.

Spiritual Healing

The term spiritual healing is a synonym for holistic healing. This is a process that involves the healing of the whole, not just a part of a person. It is also compatible with the laws of cause and effect. The foundation of spiritual healing is built on the concept of well-being. Any variation of this state of life will be initiated or caused by a statement of the mind. Holistic healing means just what it says. If you cannot heal the cause, you will never heal the resulting effect or resulting expression.

This is the foundation of the karmic laws for life on Earth. We are all here to take emotional actions, make mistakes, and learn from the result. All energies are in a polarity of expression. If we do not experience both expressions of the polarity, we eliminate the freedom of choice and the eventual determination of deciding what is our truth at that time.

During the many years of my involvement in spiritual healing, I have dealt with countless conditions of disease. I have yet to find a single case that does not involve some expression that does not partially relate to a spiritual cause for the existing condition.

For example: Supplying energy to an ulcer might heal the ulcer but the chances are that the person will develop another ulcer somewhere else. The way to eliminate this from happening would be to determine, through counseling, what could be a possible cause for the presence of the ulcer. It could be either a mental, emotional, or spiritual pattern of expressive life.

Once the possible cause has been determined, it can be identified as the cause for the creation of that condition. If the person cannot or will not acknowledge a probable cause for the ulcer, there is nothing you can do for them at that time.

In the next chapter of this book, Spiritual Causes of Disease, we have listed many expressions of disease and correlated them with what we have determined to be probable spiritual causes for these conditions. As in all matters, there are always exceptions to be considered.

A realistic approach is a necessity in dealing with illness as a whole. People are certainly susceptible to various expressions of contagious disease, such as measles, chicken pox, and mumps. These are not the conditions we are referring to in our discussion. It deserves mentioning that, even with those conditions, if the mind is in control, the severity of the contagious disease will be less severe and for a shorter period of duration.

The conditions of all forms of disability and disease that have been present from birth are exempt from this discussion. They are discussed in the chapter, Karmic Disabilities, in detail.

In most of the expressions of disease we are going to discuss, there are real medical reasons for these conditions. That must not be pushed aside or denied. What we are trying to stress to you is that, if the mind and body were in a place of truth expression, the body would be immune and have the power to remain in a healthy condition, rejecting the presence of disease.

Counseling Techniques

When a client arrives, it is your responsibility to establish the basic conditions for the healing session. Make it very clear to people that YOU are not going to cure them. Some people create the emotional illusion that after they see you, they will throw away their crutches and skip out of your office. Explain to them what healing means and what is involved in the process of becoming cured.

In creating the conditions for the session, you must create a safe atmosphere for them, as well as for yourself. Explain everything you are going to do beforehand. Do not create surprises for the client. If you do, the tension that will be created will interfere with your healing process. Tell the client that you are fully aware that they are making themselves and their body vulnerable to you by allowing you to heal them. Remind them that, as part of the process, you become vulnerable to them as well. They may not understand this so explain the following.

During the course of the healing session, you become exposed to their energies and reactions. You are the instrument for their future well-being. This could make you vulnerable to their thoughts, their emotions, and other feelings. Once they understand, tell them it is their obligation to honor your vulnerability as you honor theirs. In this way, both of you will be safe.

It is important for you to empower the client. Notify them that they are always in control of themselves. They are not there to become a victim to you. This allows them to be secure in the knowledge that they can say "no" or stop the proceedings at any time they wish to. It is important for the healing process that they do not have a sacrificial attitude during any part of the process.

This part of the counseling will set the foundation for the client to relax, feel safe, and be secure in relation to the healing process. You will have eliminated any future misunderstanding or misinterpretation of any techniques or actions on your part during the healing process.

Begin to ask the qualifying questions:

- What is the nature of your problem?
- How long has the condition existed?
- Has this condition been diagnosed by a physician and what is the result?
- Has the physician treated you for this condition?
- Are you currently under a physician's care? Do you really want and need to be well at this time?
- Are you ready to be responsible for your good health?
- If not, why do you still need to remain ill?

If you have received satisfactory answers to these questions, you are ready to proceed to explore the circumstances that could be the possible spiritual causes for the disease. First we need to look at some improper responses to the questions and recommend answers to each situation.

1. What is the nature of your problem?

"I don't know. That is what I came here for. I just don't feel well. You tell me."

Now it is time for you to inform them that you do not diagnose disease. Recommend that they see a doctor to determine if there is an existing condition that needs treatment. If they say that they do not want to go to the doctor, tell them that you cannot be of service to them without a proper diagnosis. If you weaken, feel sorry for them, or get into your ego and give them your opinion, you are out of order. It could come back and cause problems for you. How do you know that, after you give them an opinion, they may go to the doctor, tell them what you said, and ask the doctor for help?

If that happens, the doctor would be completely justified in reporting you to the proper authorities for practicing medicine without a license.

If they say that they do not believe what the doctor diagnosed and want you to feel what is wrong with them, apologize gently and tell them you are not in a legal position to make any diagnosis. You must protect yourself!

2. How long has the condition existed?

"I really don't know. It comes and goes. I don't remember when it began." This type of response tells you that the person is in avoidance. They are trying to make believe that the disease does not exist. They are either blocking their mind or ignoring it, hoping that it will just go away.

It is important to establish a time frame relating to the inception of the disease. This will open the door to this question, "What happened in your life at that time or in the course of two years before the disease expressed itself?" If they cannot recall any circumstances in that time frame, continue to go back, year by year. They will find a circumstance that created an upsetting condition in their life. It is always present. If not, they would be well.

3. Has your condition been diagnosed by a physician?

"Yes, but I wanted another opinion. I don't believe them. I want your opinion as well as theirs."

Your answer must be, "That is very wise. Consult another physician and then call me so we can discuss it further. I cannot have an opinion, as I am not medically trained to diagnose disease."

"No. I wanted to hear what you have to say first. I am afraid to go to the doctor."

This is a very sensitive situation. My advice is to just smile, say nothing, and proceed to give them an energy examination. If you feel any energy distortion, heat disturbance say the following to them. "I am feeling some heat in this area of your body. I don't know what it is but I encourage and support you in going to the doctor. It could be nothing at all, but make sure."

Doing this makes them feel supported and could give them the courage to see their doctor for an examination.

4. Has a physician treated you for this condition?

"Yes, but I don't seem to be getting any better." "No, I don't believe in doctors." "No, I would rather be healed spiritually."

This is the moment to begin counseling. They have just intimated to you that they are either unaware of, or have shut out, any cause of their condition. They do not want to be responsible. They want either the doctor or you to bear the responsibility if they recover or remain ill.

Their statements free you to accept them for healing. However, tell them that if they decide to see a doctor, please inform you and your treatments must end. They need to understand that you are only involved in a spiritual process. You cannot guarantee anything or make indications that could lead to false hopes or securities on their part.

5. Are you currently under a physician's care?

"Yes, but I want healings as well."

Ask them if they have discussed this with their physician. If they have and the doctor is open to spiritual healing, have them bring a note from the doctor that acknowledges permission and approval for the healing sessions. If they have not consulted about this with their doctor, tell them to do so. If their response is that their doctor would not approve, it is appropriate for you to tell them that you cannot treat them while they are under the doctor's care.

They could respond, "I won't tell the doctor that I am coming here for healings." Stay in your integrity! Do not place yourself at any risk.

6. Do you need and want to be well at this time?

"I think so." "I never thought about it." "My friend told me I should do something about this."

These responses give you invaluable clues to the nature of your counseling sessions. I advise you not to proceed with any healing until the person has responded in a positive way to your question.

If you feel that the problems are deep-rooted and you are not qualified to counsel them, send them to someone with the proper credentials for this situation. Do not experiment. You are dealing with a person's life. If you receive this kind of response, you are dealing with a very serious condition. The person is trying to live a real life in the expression of illusion. They are in the avoidance of life and could become vulnerable to self-destructive energies in the future.

7. Are you ready to accept the responsibility of good health at this time?

"I don't understand what you mean?" "I don't know."

The nature of these responses opens the door for you to begin to make them aware of the power and effect of the thoughts from the mind. From this point, you begin counseling, stressing the freedom to make mistakes in life. You are probably dealing with someone who tries to be a perfectionist. This is one of the heaviest burdens of life and denies success repeatedly, again and again. If their answer is not positive, ask them:

8. Why do you still need the disease?

"People notice me." "It brings me attention." "It doesn't matter."

I could give you many more statements of unworthiness. Each gives you a clue to examine, discuss, and bring out from the depths of their subconsciousness.

As you commence with your counseling, your first goal is to determine what could be a possible spiritual cause for the disease. You do not have to exactly pinpoint the cause, only to have them consider that it is a possibility. This allows them to be open and receptive to the healing energies. Remember, if the person is not open to be healed, it will not be effective. Their mind will reject the energy and it will just pass out of the body.

In the course of my healing service, I have determined that the main spiritual causes for disease fall into specific general categories.

1. Dysfunctional environmental upbringing.
2. The absence of the experience of success.
3. Sexual rejection and unworthiness.
4. Self-judgment.
5. Prolonged anger or resentment against yourself or others.
6. The absence of the experience of love.

1. Dysfunctional Environmental Upbringing.

This portion of the energy patterns that become integrated with the sub-conscious responses is mainly associated with the family structure and environment. This includes:

a) Any form of abuse, either physical, emotional, mental, sexual, or spiritual.
b) The absence of support towards fulfilling personal goals.
c) The rejection of personal expression. Imposed silence by parents.
d) Isolation: Either voluntary or imposed as punishment.
e) Comparisons with siblings.
f) Absence of parental nurturing.

2. The absence of the experience of success.

When a soul incarnates in this society, it becomes conditioned to the patterns of the society of life relating to success and failure. When we, as part of that society, do not experience personal success, success that relates to us as individuals, we consider ourselves to be failures. In time, a response program could develop in the sub-conscious mind that says we are not worthy or good enough to have success.

In your counseling, try to stress that there is no real condition that can be labeled failure. There are only different degrees of success. What is considered to be success is always relative to the individual's perception. What one person considers successful could be labeled failure by someone else.

The most powerful obstacle is to consider yourself to be a perfectionist. This is a false label and a false goal. In reality, anyone who claims to be a perfectionist never allows themselves to be good enough. They always fail, believing that they could do better and better.

They need to begin to acknowledge their efforts, then learn from their mistakes, and be more successful the next time. Success is the motivation and incentive for continual growth and development.

Suggest to your client a series of actions that will initiate the process of creating small success for them, one after another. It does not matter how minor the success is. The importance is in the emotional result of the experience. In time, the pre-conditioned pattern of failure will be replaced with patterns of achievement. This will create a goal-oriented and achieving client.

3.   Sexual unworthiness.

This destructive reaction is created from the lack of understanding of the proper role or function during the intimacy of sharing in a physical, sexual interaction. Reassure them that they do not have an assigned role in the action of intimacy. It is not to be an experience of giving themselves in sacrifice or becoming a victim by the use of force. Physical intimacy is an equal sharing between two people for the benefit and pleasure of both of them, not only for one or the other. The old attitude of dominant and submissive roles belongs to the days of slavery. We all have the great power of free-will. We all have the ability to say the wonderful words, "No, thank you", and honor our body and our soul.

4.   Self-judgment.

When a person judges themselves, they are saying, "I am not worthy of my soul and of God." The ultimate cause for this lack of acceptance that brings out the expression of unworthiness is the inability to activate Grace in that person's life.

When we can begin to accept the pulsating polarities of universal energy, we conduct our actions and our lives knowing that everything in the Universe, is in a constant state of change. The only constant realities are God and soul. If a person can begin to accept these words, they can free themselves of what is commonly known as the straight line of life.

We often refer to this as the shortest distance between two points. Imagine walking on a country lane unable to see anything on either side, only the road in front of you.
You would miss the beauty of nature, the people waving to you as you pass by, and the chance to wander through a forest.

Life is an undulating, pulsating, curving spiral of Light. Walk in this Light! Allow it to carry you through the waters that are not always smooth or the roads that sometimes have mountains in your path. If you can accept the heartbeat of life, you can begin to live in Grace without self-judgment.

5. Prolonged anger or resentment against yourself or others.

Either of these expressions is always a condition of judgment. The longer you hold on to the judgment, the more powerful it becomes and the more distortion you have in your life.

Universal Law states: As you judge yourself or others, so it shall return to you increased tenfold.
This expression is the singular, greatest cause for most expressions of disease. Explain to the client that we do not really become angry with other people. We become angry with ourselves for allowing and accepting involvement in a particular situation. What we do is to become aggressive towards them and we vent the anger at them but that is really expressing our frustration with ourselves.

If you can explain the Law of Allowance to them, it will help them to let go and let live. (See section on Allowance in this chapter)

6. The absence of the experience of love.

Begin by explaining the distinction between physical and spiritual love. This lack of understanding is responsible for many problems in relationships and other areas of life.

Physical love is limited to emotional and sexual energies. This can and karmically is supposed to create insecurities, sacrifice, and the uncertainties of life.

Spiritual love is the experiencing and sharing of soul energies using the physical body as the vehicle to complete the sharing process.

We all have the tendency to blame ourselves if someone does not return our feelings and express them to us. We feel that there must be something wrong with us, not them. You treat this program by creating an experience of self-love inside of your client for themselves.

This will be a profound experience for them. It will begin to negate much of the past programming of self-blame and unworthiness. The percentage of people who walk through life believing that they will never feel love or cannot express love is much greater then you can imagine. By creating this self-experience for them, their old program rapidly becomes healed.
Allowance

When you have reached the place in your counseling where the client is ready to let go in love or express allowance, a substantial portion of the healing will have already been accomplished.

Many times I ask a client, "How much longer do you intend to give this person power and control over you?" This statement usually gets them to begin to think and, hopefully, a light will go on inside of their mind. They do not realize that because they were angry or resentful towards someone, they gave away part of their own personal power and control of their life.

Learning the distinction between love and allowance brings forth a greater understanding of the proper way to express a relationship. One that is casual or of an intimate nature.

All bibles tell us to Love thy neighbor as thyself. Have we interpreted this to mean that no matter what people do to us, we are to turn the other cheek and love them?

The subtle message behind this spells sacrifice in large letters. Are we supposed to love everyone regardless of what they do to us? The more you think about this, the more you need to look inside for the true meaning of these words.

You find them in the Law of Allowance. When you allow your neighbors as yourself, you are truly loving them. You become free. You own your free-will and all the power and control of your life.

The Law of Allowance states: You will not judge or make any determination of the actions or statements of others. You will freely allow all to own the responsibilities of their actions and for the conduct of their lives.

Practicing allowance requires understanding the difference between decision and judgment. Without this understanding, it is difficult to comprehend the proper attitude needed for complete freedom.

A decision is a statement of mind that is reached based on the reaction to a prior action that was proposed or taken by someone else. That action is either compatible with your truth or it is not. The decision may come as the result of a single action or one that is continually repeated.

A judgment is created by the voluntary, unsolicited repeating of your decision to other people. This can pertain to a person, a thing, a place, or other expression of life.

For example: John and Jane begin a relationship. John decides that they are not compatible with each other. He tells Jane that they cannot continue the relationship because of the difference in their perception of life. Jane could say, "Who are you to judge me?" These words would not be valid. They are not a judgment. They are John's decision. He has a right to make a decision.

If, however, after ending the relationship, John begins to call other people and complain, making unsolicited statements, his decision has changed into an expression of judgment.

The line between decision and judgment is often very fine. The best way to avoid crossing the line into judgment is to remain in allowance at all times. It is not a judgment to approach someone and say, "I find that I do not agree with your actions or your way of life.

I respect your right to be who you are but I must end our association." This is a pure decision keeping you in allowance.

The most effective way to remain in allowance is to eliminate the word that causes so much pain and disappointment in life.
The word is expectation. This is an unstable, destructive emotion that creates total vulnerability.

Expectations are silent judgments.
- I expect them to do it like I do it.
- I expect them to change.
- I expect them to understand me.
- I expect them to accept, like, love, etc., me.

The Law of Allowance opens the door for the acceptance of the present expression. If you cannot accept that, look for another relationship, otherwise, you will move right into judgments and begin to give it some of your power.

A client of mine called me and announced that she had just become married. I am going to quote her words. "Hi Frank, I just got married. He's a nice guy but there are a lot of things I have to get him to change. I'll straighten him out." I laughed inside, for what she really said was, "I expect him to change. I will not allow him to remain as he wants to be." What a disaster! She had set herself up for the return of judgment energies that could cause her problems later on in life.

Using this principle during your healing sessions will greatly assist you in helping your client to cut the cord to past attachments enabling them to move forward into a more free responsible expression of their life.

My goodness, who would believe that there are so many things to discover and make decisions about? Maybe it is too much trouble and too complicated. Maybe it is better just to float along dealing with life as it shows up. No! It isn't enough. You need to understand yourself. You need to begin a little self-analysis and find out why you have not been successful in the past. What has happened?

Have you called something a failure because you have not understood who you are or what your real needs are?

It is important to answer all these questions and acknowledge the answers to yourself. Slowly, you will begin to understand who you are. A wonderful, imperfect human being! As your path of self-discovery unfolds, you will be able to begin to have the capacity to understand others as they present themselves to you.

To truly understand someone, you need to look at them through their eyes, not yours, and allow them to be who they have chosen to be.

Holistic Healing Procedure

By definition, holistic healing means that the practitioner is concerned with the complete person, not only what seems to be the current expression of disease. Treating a physical manifestation does not eliminate the cause or need for the existence of the disease. It is for this reason that every spiritual healing must be preceded by a period of counseling. This will assist in the determination of what could be the possible mind-cause for the presence of the disease.

I am going to describe several circumstances that I encountered over the years that I have been involved with in the field of spiritual healing.

The client arrived and we sat and chatted for a few minutes. I asked her what was her problem and why had she come to see me. She replied, "I am not going to tell you what is wrong with me. You are the healer. You tell me."

I smiled at her and politely told her that there was nothing I could do for her and perhaps she should go to see someone else. She did not understand, so I explained the following to her. I told her that I was not a licensed doctor. Therefore, I was not qualified to diagnose any expression of disease. I also said that I was not going to play any psychic games with her so she could see if I was a good psychic or not.

Those words are very important for you to remember. As a spiritual healer, you must avoid the pitfalls of your ego and not play doctor. You must remember your role of service and its limitations. If you diagnose disease, you can be placed in jail for practicing medicine without a license. The truth is, you would deserve to be in jail. Please remain in your integrity at all times. You serve people with a "no" when it is appropriate and applies to the situation.
Many people will come to test you. They want to see if you can discover their problem, without their help, just for fun. Do not allow yourself to be compromised in this manner. Serve people who come to see you in sincerity and in honesty.

I asked the client if she was currently under the care of a doctor. This must be the first question you ask anyone who comes to you for healing. She said, "Yes, but I want to use everything to help me recover." My answer was that I could not treat her while she was under the care of a doctor, unless she obtained written permission from the doctor for me to treat her.

If I had administered healing under those conditions, I would have been interfering with the practice of medicine unless she had obtained written permission from the doctor for me to work with her.

I have related these circumstances to you in order to clarify certain conditions that are not to be compromised under any conditions.

1. **Never diagnose disease unless you are a legal physician.**

2. **Do not allow your ego to push you out of integrity and play doctor.**

3. **Never treat anyone who is under the care of a physician without the written permission and approval of the doctor.**

4. **Do not become involved in situations that you are not qualified to treat.**

5. **Do not massage or manipulate the body unless you are legally certified.**

Example

A friend asked me to look at her mother. It seemed that she has visited a healer, who told her that she had lung cancer. This was a statement from the healer that did not relate to the reason she came to see him. Naturally, my friend took her mother to see three different cancer specialists. Each one told her that there was no trace of cancerous tissue in her body. She was very healthy.

After each examination by a doctor, the woman just shook her head and said, "The healer told me I have cancer."

I checked her energy patterns and immune system, and tried to reassure her that there was nothing wrong with her. My words fell on deaf ears. Her mind was totally closed to my words.

The sad part of this story is that the woman could have contracted lung cancer, as she had accepted the condition in her mind. She had resigned herself and her body to the presence of this disease.

I hope that this example of the violation of integrity and the lack of responsibility to the client will serve you in creating the awareness of the great responsibility you assume as a spiritual healer. If you will always remain within the guidelines of your service as an instrument for God, you will maintain your integrity and never be out of order.

The Chakras

A chakra is a physically invisible, spiritual energy receptacle or opening, through which energy enters into the network of meridians present in the physical body. They are symbolically represented as closed, multi-petalled flowers that open as the individual begins to recognize and become aware of their relationship and connection to their soul, and to God.

When the petals are totally open, they serve as a collecting dish to receive God's Light and disseminate it throughout the energy meridians of the physical body.

The basic chakra system is divided into several functional categories. The primary group of chakras are the seven basic operational centers for the current karmic, physical incarnation. They correlate with the karmic laws of progression that are assigned to the conduct of life on our planet Earth. These chakras are sub-divided into three distinct energy categories relating to the specific karmic functions for life.

1. The higher chakras
2. The lower chakras
3. The heart chakra

The Higher Chakras

These centers serve for the growth, perception, reception and expression of the spiritual aspects of life. They receive frequencies of energy that stimulate the God-Centers of expression. This enables the individual to begin their path of intuitive perception and the awakening of the dormant areas of the mind to the presence of their soul.
Before a person has spiritually awakened or is on their conscious path of growth, the location of these chakras are in the form of a triangular pattern of energy.

1. The Crown chakra is located at a center point on the head just above the hairline.

2. The Third Eye is situated one inch, or two centimeters above, and between the eyebrows. At this time, the center is one inch, or two centimeters to the left of center, and remains there until the individual begins their spiritual awareness.

3. The Throat chakra is located in the lower portion of the throat, just below the Adam's apple.

When a person commences their spiritual growth and attitude, as a part of their expression, the Third Eye begins to shift into its permanent position in between the physical eyes. This process symbolically opens the chakra and enables the person to begin to see into the energies of their soul.

In the earlier years of life, the off-center position of the Third Eye allows people to have their emotional, karmic experiences without using their intuition. They take their actions from emotional decisions to make mistakes and have the choice of the polarity of expression.

The Crown chakra represents a spiritual antenna — a receiver and absorber of energy frequencies that nourishes your God-self, feeds your sub-conscious mind with knowledge and energizes your physical structure. It is through the Crown chakra that you receive and experience your connection to the Creator-God. This will begin the creation of inner feelings of worthiness and commence the process of activating your intuitive processes.

The primary function of the Third Eye is to project spiritual vision. As long as a person is spiritually asleep, this chakra will remain closed. This will result in creating the condition where a person's insight or vision is limited to physical and emotional receptive centers and reactions. The energy rods and cones behind the eyes, that activate spiritual and energy perception, will continue to remain dormant.

It has been normal for this condition to exist for the first five cycles, or thirty-five years of life. During that time, most decisions, wants and needs are determined through the emotional, physical eyes.

If spiritual growth becomes activated, the dormant rods and cones become stimulated, and the perception of energy, thought, light, and color received through the Third Eye will begin to express themselves in the corresponding centers of the brain. This will become transferred into conscious mind awareness.

When the Third Eye chakra has moved into center, the three Higher chakras come into alignment, and create the upper half of the body's immune system of spiritual and energy power in connection to the soul.

As growth continues, the chakra gradually moves to the center of the forehead between the eyes. This forms a visual triangle between the third eye and the two physical eyes, allowing all perception to be experienced in balance. The person can begin to see things as they truly are, not as they only seem to be through emotions alone.

The Throat chakra is the safety valve of the expression of all energies. Many times our throat warns us that we have something that we are holding inside and not expressing. Our throat closes, we cough, and we feel like something is stuck inside. We develop throat cancer when all we express is sacrifice and lies.
All communication energies travel through the body to be released through the throat. The power of our breath can serve to create the courage for our expression.

The Lower Chakras

We refer to the lower chakras as the Earth centers of energy that create the balance of our integrated expression. One of the most powerful statements we can make to you is, "Your soul belongs to God, and your body belongs to Mother Earth." When the time comes that you begin to correlate both of these sources of energy, and use them in your daily life as well as your healing service, you present an integrated expression of balance and well-being.

The Solar Plexus chakra is our point of vulnerability. This is the entry place for most energy that we take in, or allow to be projected into our body.

It is the center that retains reactive energy of all sacrifice actions and guilt responses. How many times have you felt a pain in that area when you are in a strange or uncomfortable situation? When you are being sympathetic to someone, you often have an inner reaction to their situation, either of emotions or a physical symptom of mutual suffering.

The Second chakra is the emotional memory reactive center for all our responses to every expression of dysfunctional behavior. This includes all categories of abuse, and the positive and negative programming of all actions that involve our exchange with the people in our lives. It is from the energy memory of the Second chakra that we develop all the negative affirmations that affect our health, as well as the incentive energy to move forward in a successful life. This chakra is the major energy receptacle that causes the negative responses to self-Grace. It has the most powerful effect on our conscious behavior, as we move through life trying to discover our true identity.

In the earlier years of life, the Second chakra is also located to the left of center. This creates the Earth triangle, and is significantly important in the stimulation of moving people into their emotional, free-will expression. As the years of time and experience begin to stimulate spiritual awareness, the chakra moves into center, creating alignment of the Lower chakras.

The Base chakra is a functional center of expression. It does not store any energy reactions created from an experience. These reactive energies are stored in the Second chakra.
The most intense and powerful experience of life on Earth is to discover, learn, and express what is called love. This lack of understanding and proper expression is the greatest cause of many types of dysfunctional behavior. This is the true function of the Base chakra. We need to find the answer to this question. "Is our Base chakra the expression of our emotional, sexual behavior, or is it the tool of our truth expression of our soul?"

The Heart Chakra

The Heart chakra is the connecting balance between the Higher and Lower chakras. It controls the dissemination of soul energies to all areas of the body. We sometimes call the Heart chakra the house, or garden of the soul. It is in this chakra that the core or nucleus of the soul exists. The core never leaves. If it does, we have died. From this position, the soul can emit projections of its energy to all areas of the conscious and subconscious expressions. This is how the soul attempts to correlate the life expression with its purposes for incarnation.

With the balance of the Higher and Lower chakras in place, the immune system functions properly to maintain the health and balance of the body and the mind. As long as we remain in the conscious expression of our soul truth, health is ours. Whenever we take an action that is not our real truth and in compatibility with our soul, the related chakra shifts out of alignment. The result is a weakening of the flow of energy in the center meridian, or immune system, making us susceptible to disease.

These are the seven basic chakras or energy centers in the physical body. There are many more chakras and, as time passes and people evolve, we continue to become aware of their presence and proper functions in life and health.

I am going to describe twelve additional chakras located on the front of the body and twelve chakras located on the back of the body. The activation and use of these centers greatly enhances your healing efficiency, as well as contributing to the maintenance of good health and spiritual development.

Front Chakras

The magnetic third eye

This chakra is located exactly between the Crown chakra and the Third Eye chakra in the center of the forehead. It functions with magnetic energy. When it becomes activated, it serves as a powerful antenna for receiving channeling and magnetic energies from universal levels. These energies can be used for healing and outer dimensional communication.

Underarm chakras

They are located under each arm, where the arm joins the body. They are used to energize energy blocks in the upper arms and shoulders. They are used in conjunction with energy insertion points along the top of the shoulders for treating bursitis or arthritis in the shoulder area.

Inside elbow chakras

Located in the inside bend of the elbow on each arm, these chakras energize conditions that affect the elbows, such as tennis elbow and arthritis. When they are used in connection with the underarm chakras, they energize the upper arm area as well.

Inner wrist chakras

The chakras are located on the inside joint of the wrists. They are the lower arm connections to the underarm chakras for energizing the entire arm. They also serve for carpal tunnel conditions, and arthritis in the wrists and fingers.

The magnetic birth chakra

The chakra is located at the navel. It is a spiritual umbilical cord to the outer dimensional energy of what is called the Universal Central Sun, a symbolic area of the universe that births the evolutionary paths of newly created souls. This is used for healing, when you are dealing with a person who has doubts about the existence of the Creator-God.

Hip chakras

The chakras are on the top side of each hipbone. They are used as the upper connection for energizing the entire leg for circulation conditions as well as any circumstance of hip disease.

Ankle chakras

They are located on the inside of each anklebone. Their function correlates with the chakras on the bottom of the feet in relation to all ankle, foot, and toe conditions.

Back Chakras

The clairvoyance chakra

This chakra is not used for physical healing. When energized, it increases the capacity for the development to sense and feel energy patterns in other people and places.

The channeling chakra

It is located at the base of the skull on the spine. When energized, the increase of energy assists in the development of strengthening the connection to energy transmission from spirit levels of existence.

Upper thigh chakras

The chakras are on the upper part of each thigh where they meet the buttocks. These are very powerful and important chakras. By energizing them together, they create a flow of energy through the base chakra. If someone has been a victim of sexual abuse, this energizing will help to re-open the base chakra, place love vibrations inside, and begin to create a new program of positive reactions to the body center. This treatment has proven to be very effective in the healing of sexual abused people.

Knee chakras

They are located behind the knee where it bends. They are used for all knee conditions of disease. Many times they are used in conjunction with the energy insertion points located on each side of the knee. This makes a triangulation of energy, which is very effective.

Feet chakras

The chakras are situated on the bottom of each foot, in the center of the fleshy part of the ball of the foot. It connects the leg circuit from the hip chakra to the toes. It also energizes the reflexology points in the foot and sends the energy to the associated organs of the body.

On the back of the body torso are located four additional chakras. We are grouping them separately for specific reasons. These chakras are high energy chakras used for healing conditions of advanced disease.

Before a person has begun to spiritually evolve, and accept themselves as an extension of their soul, these chakras are not open or available to be used. This is important to remember. If you try to use them under other conditions, nothing will happen. You will be wasting energy. There are no exceptions.

Higher base chakra

The chakra is located in the spine, in the lower small of the back, just above the buttocks. It corresponds to the normal base chakra, but is used for serious disease affecting the genital area and its organs.

Higher second chakra

This chakra is also on the spine approximately four inches, or nine centimeters above the higher base chakra. It functions in the same capacity as the second chakra, with greatly increased energy intensity. In addition, it is to be used for advanced disease of all organs in the lower part of the body.

Shoulder blade chakras

The location is on the center of each shoulder blade on the back. They are always used together. They are not connected to any chakras on the front of the body. We use these chakras for all conditions of advanced disease in the lung and chest areas of the body, including the breast tissue.

The use of the chakras in energy healing creates a direct entry into the meridian system of the body. As the person accepts, and increases their awareness and belief in the spiritual part of their life, the chakras slowly unfold wider and wider. This means growth. This creates a continuing sensitivity to life and to soul existence.

When using a chakra to heal, you have a mind-thought and open the chakra to yourself. These are important words for you to express. They protect your client from other energies entering their body. When you are finished with the healing, make sure to have another thought: I close the chakra back to its normal position.

If the chakra is left open wider than normal, energy will continue to pour inside, and could have a disturbing effect on the person. The reaction might be of real physical discomfort and pain from the overdose of energy entering the body.

The Chakra System (front)

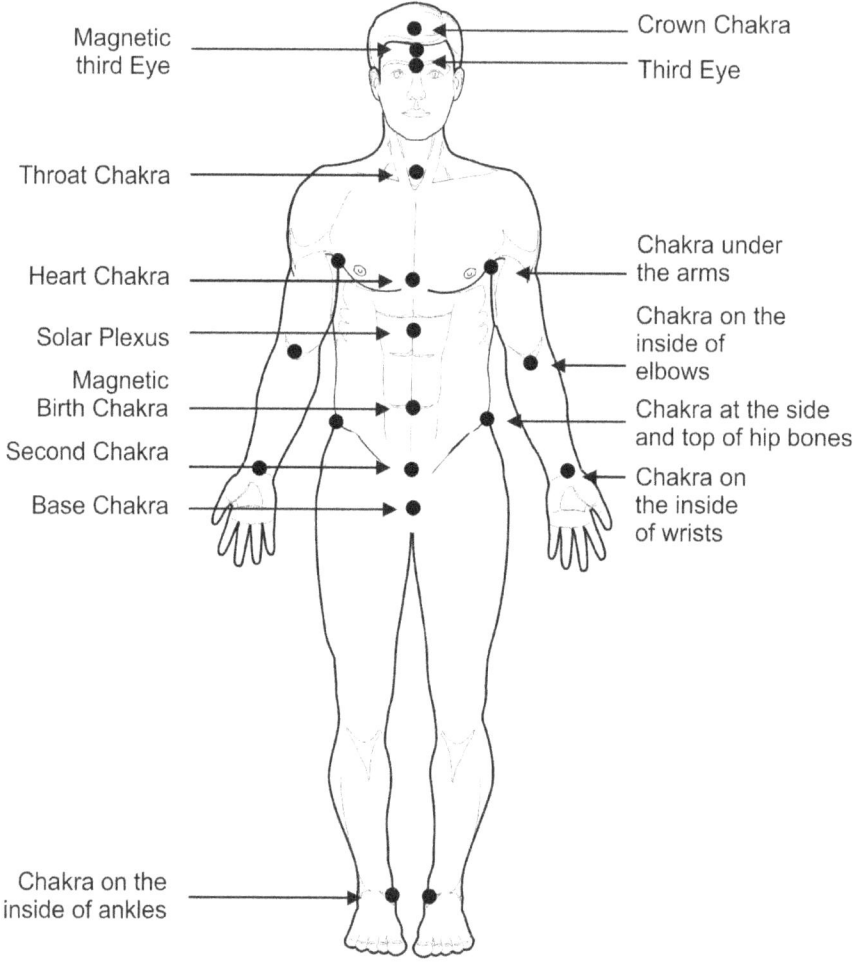

The Chakra System (back)

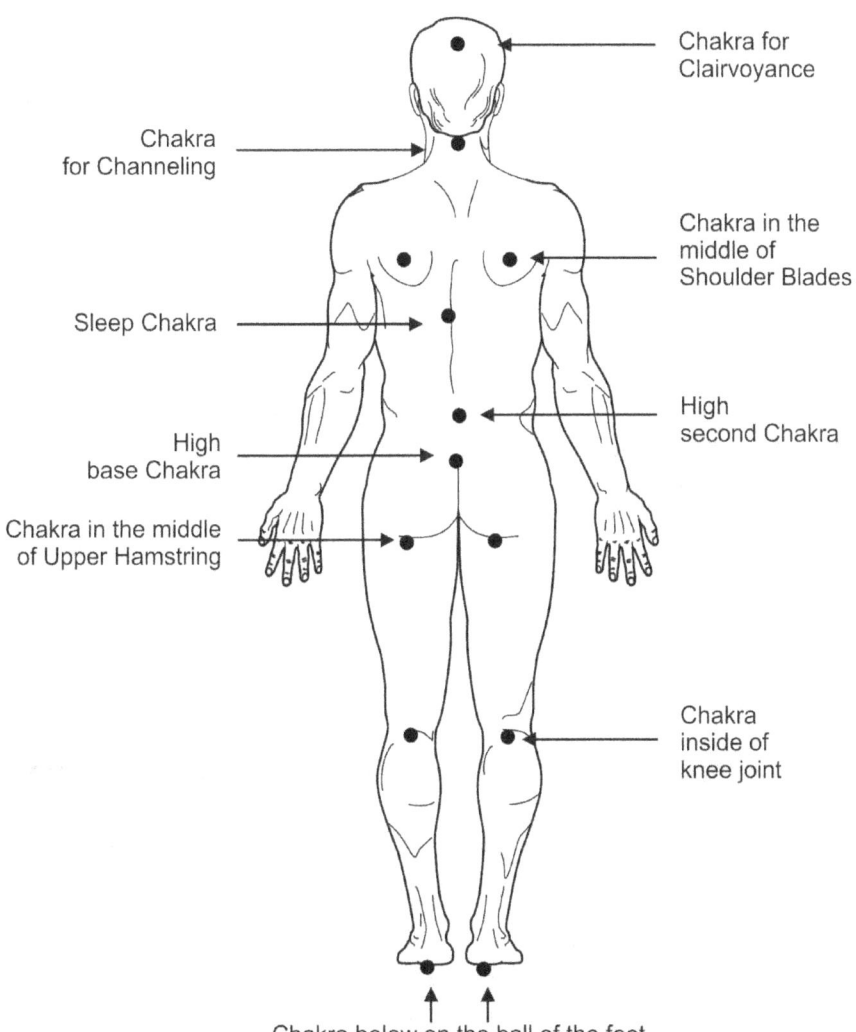

Energy Insertion Points

Energy insertion points are specific places on the surface of the body that serve as entry points into the energy meridians of the body. These points are not to be considered as chakras. They do not have a spiritual construction or connection to the God-source.

The physical body has many hundreds of places through which energy can enter or be placed inside. In the application of acupuncture the practitioner uses any of two thousand entry points for the insertion of the needles.

We are going to describe many of the most commonly used insertion points on the body that we use for healing energies. We encourage you to practice your energy healing through these points. This will begin to build your confidence in your ability to transmit energy to others in an effective manner.

It is possible to attempt insertion point healing on yourself. You will find it far more effective if applied by another person. We all have the tendency to get in our own way. That means we have created the condition and it is most difficult to place our feelings aside and be objective with ourselves.

1. On each side of the head, one inch or two centimeters directly above the top of the ears.

Purpose: To reconnect energy short circuits in the brain that pertains to seizures, headaches and lapses of mental functions.

Application: Place the middle finger of each hand on the insertion point. Project a mind thought to send energy from one finger through the head to the other finger. This will generate a circuit of energy to restore a normal flow of energy between the two insertion points. If the circuit is restored, you will feel a gentle pulse in each of your fingertips. Do not apply this application for more than two minutes at a time. If necessary, the healing may be repeated every two hours.

2. On each side of the face, one inch or two centimeters back from the corner of each eye.

Purpose: To create a circuitry that will supply energy to the eyes for conditions of eyestrain, myopia, astigmatism, lazy eyes and dyslexia.

Application: Place the middle finger of each hand on an insertion point. Mentally project the thought to send energy to the eyes for the specific existing condition. This application should be for a duration of only two minutes and may be repeated every two hours, three or four times daily. This technique of healing is not for conditions of glaucoma, cataracts or corneal problems.

3. On each side of the head just in back of each ear, and in the hollow even with the bottom of the ear.

Purpose: To supply energy for most conditions of disease pertaining to the ears, nose and throat.

Application: Place the first and middle finger of each hand on the proper insertion points. Mentally state the purpose for the healing and the area to receive the supply of energy you are transmitting. Maintain this position for no longer than two minutes. The application can be repeated every two hours.

4. Below each eye, on the sinus cavities.

Purpose: To reduce swelling, infection, and congestion in the sinus cavities.

Application: Gently place the middle finger of each hand directly on the sinus cavity on both sides of the nose. Mentally state that you are sending heat energy into the sinus cavities. This may serve to assist in loosening the existing congestion. The position can be maintained for five minutes and repeated in several hours.

5. The empathic center.

This insertion point is located on the exact center of the top of the head.

If this center is open, you can unknowingly experience pain from other people and transfer it to yourself and your body. You are able to pick up their thoughts as well as their emotions. Many times this center will open by itself and create these conditions that are not desirable for you.

Purpose: To close the Empathic center and free the person from the disturbing influences of other people. Once the center is closed, the person can open it to themselves with the mental statement, "I open my Empathic center to myself." When they have finished using this intuitive center, they should make another statement, "I close my Empathic center." This will protect them from any further invasion of outside energies.

Application: Place the fingertips of one hand flat on the exact location and make the following statement. I send energy to close the Empathic center from outside energies. Keep your fingers in place for several minutes to complete the energizing. Tell the person that they may open this center by means of the thought: I open my Empathic center to myself. When they have finished using the center, they are to close it with another thought. I close my Empathic center. The person will feel it closing and will be protected once again.

6. Complete body energizing.

Purpose: This configuration connects your fingers, in an ancient pattern, to the main energy endings in your head. It is used to energize the complete physical body, from the head to the feet. This healing is used for all conditions of fatigue, general low energy conditions, emotional stress and any expression of disease that can affect the whole body structure. Some of the conditions of disease include high or low blood pressure, poor circulation, allergies and most expressions of skin eruptions.

Application: Place both thumbs on the Clairvoyant chakra located at the back top of the head. Set both index fingers on the Crown chakra located just behind the hairline. This forms a triangle. Arch your wrists in a relaxed manner and allow the remaining fingers to spread out in a fan pattern resting on the side curve of the head. Set the condition in your mind to send energy through your fingers into the body meridian structure to energize the entire body.

Remain in this position for a full five minutes. Have the person remain quiet for an additional five minutes to allow the body to adjust to the infusion of energy. The treatment may be repeated on a daily basis for as long as necessary.

7. Magnetic ear healing.

Purpose: To insert magnetic energy into the ears for the purpose of energizing all conditions that affect the ears, nose, and throat areas.

Application: Hold your hands in a perpendicular position. Bend your fourth and fifth finger against the palm of your hand, as they are not involved in this process. Spread the first and middle fingers into a V position and place your thumb between the fingers. You have now created a configuration that will create a vortex of energy. Place your hands directly in front of the ear openings, positioning them four to five inches away from the head (nine to ten centimeters). From your mind, send energy down the length of your arms and out of the thumbs into the V you have formed.

This application generates an extremely powerful force field of energy. Under no conditions should you maintain this position for more than two minutes. If held for a longer period of time, dizziness or disorientation could result. Once again, allow the person to rest for five minutes.

8. Shoulder insertion points

These points are located along the top of the shoulders. Four on each side of the body.

Purpose: To initiate the insertion of energy to all areas of the arms and upper body torso.

Applications:

a) Place the palms of both hands on the shoulders. Mentally, send energy down into the upper half of the body. This will supply fresh power to the areas that may be in a weakened energy condition.

b) Place several fingers of a hand on a shoulder. You will insert energy into the body through these fingers. Place the other hand on the area of the body that needs the energy.

c) From your mind, insert the energy from the shoulder position down to your other hand and remove the tired energy to the surface of your hand. This is important. You do not want that energy to enter your body.

d) This application involves sending energy to assist with infection and other weakened energy conditions in the upper half of the body.

9. Insertion points in the palms of the hands.

Purpose: These energy insertion points have a dual function. They disseminate energy from yourself to other people as well as taking in energy for your own well-being. The points are utilized for all conditions or energy blocks located in the arms, from the shoulders to the fingertips.

Application: You will use these insertion points to remove tired energy from the arms. The energy will be inserted from a higher point. The shoulders, elbows or wrists, and will be removed by the mind thought through the insertion point in the hand.

When dealing with the fingers, as in arthritis, the energy will be inserted through the palms and removed through the fingertips.

10. The ten fingertips of the hands.

Purpose: These insertion points serve to transmit as well as to receive energy. Many healing applications, in addition to energizing, are conducted through the fingertips. The prime function is to assist in removing energy blocks existing in the joints of the fingers. The healing process often involves the dissemination of energy through the fingers to many areas of the body. As a person performs continual healing with the use of their hands, the fingertips develop a strong sensitivity to the condition of energy in a client's body.

Application: Energy is inserted in the center of the client's palm with several fingers. Place the fingers of your other hand on the client's fingertips and mentally send energy into the palm of their hand.

Pull out the tired energy with the fingers of your hand that touch their fingertips. This process may greatly assist in removing any energy blocks that have formed in the joints of all the fingers.

11. Insertion points between thumbs and first fingers.

Purpose: These insertion points are called, The Gateway to the Soul. By creating the proper conditions, two people can create an emotional and spiritual bonding of peace and love between them. Many times this energy exchange will eliminate stress and discomfort in a relationship, by the creation of the conscious awareness of experiencing each other's soul energy.

Application: One person extends their hands with their palms up. The other person extends their hands with the palms down. Each person places the tip of their thumbs between the thumb and first finger of each hand. This establishes the energy circuit.
Close your eyes and just begin to experience the exchange of energy that begins to take place. Keep your mind relaxed and the love healing will begin to flow inside both of you. Thoughts may begin to enter your mind. Allow them to be expressed and share them with each other. This will serve to further establish a closer connection of harmony and balance between the two individuals.

12. Insertion points on the center, top of the thighs.

Purpose: The main purpose for this category of healing is to remove and heal emotional and sexual blocks. It is also very effective in replacing energy relating to prostate conditions, vaginal infections, and menstrual cycle regularity.

Application: Place two fingers on the top of each thigh where it meets the trunk of the body and mentally project energy into the insertion points. Your goal is to create an energy circuit between both thighs that moves through the genital and emotional expressive center of the body. Hold this position for five minutes and then release. During this process, the person will begin to feel sensations in this area. They may initially react in discomfort or fear.

Keep talking to them and reassuring them to continue and to move through the experience to a new level — one that is not sexual, but spiritual and secure. In several minutes they should begin to experience warmth, peace, and experience comfort and security with the new experience of themselves in relation to their body.

13. Insertion points on each side of both knees.

Purpose: To work with all energy conditions that can affect the normal functions of the knee.

Application: Place two fingers of each hand on both sides of the middle of one knee. Mentally send energy from one hand to the other. This will create a circuit of energy that will flow back and forth between your hands. Continue this application for five minutes. The creation of a circuitry will help to break down any existing block in the knee that may have caused stiffness, swelling and soreness in the knee area. This healing can be most effective when a condition of arthritis is present in the knee joints.

14. Insertion points at the tip of each of the toes.

Purpose: To alleviate pain and energy blocks that may have formed in the toes of the feet.

Application: Place the thumb of one hand on the chakra located on the ball of the foot. Place the fingers of your other hand on the tips of the toes on that foot. Mentally send energy into the chakra on the ball of the foot and direct it to your fingers at the tips of the toes. Maintain this position for five minutes. During this process, the person may experience slight pain. If this occurs, tell them it is the result of the dissolving energy blocks and will dissipate within several minutes.

15. Energy insertion point over the physical heart.

Purpose: To ease tension, emotional distress and to allow the person to experience love in a safe and secure manner.

Application: Place one hand flat on top of the physical heart. Draw energy into yourself through your Crown chakra.

Close your eyes, relax, and send the energy through your hand into the person's heart. You are sending love. Not emotional love, but the love of Grace, God's love. If a person has any expression of heart condition, do not place your hand on their body. Keep your hand at least six inches or fifteen centimeters above the body. Set the condition in your mind, that you are sending a soft, gentle vibration through the energy insertion point in a frequency that is compatible with their hearts energy field.

These are the main energy insertion points to use for inserting healing energy into the body. Most of them can be used in conjunction with the chakra located near the area to be healed.

Energy Insertion Points

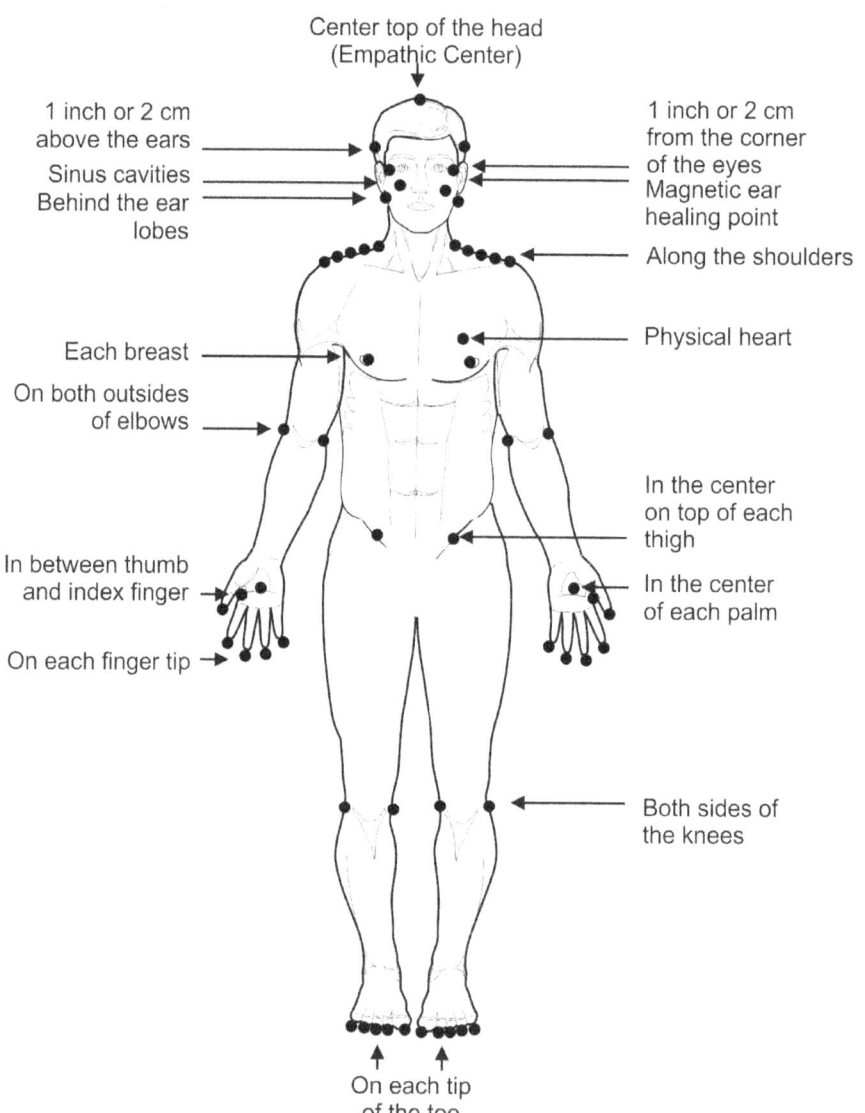

Life Is

I am going to relate a channeling to you that bears a most important relationship to our life here on Earth. It tells the story of the function and purpose for the karmic, life experiences on planet Earth. When you read these words, open your mind and move inside of the words to begin to feel, what is the relationship to your life and to your past experiences. Assimilating these concepts will give you a greater insight into people and be of great assistance to you in your counseling and healing services.

We are the "Original Light." One might ask, what was before the Original Light? It normally would be a most difficult, almost impossible question to answer but there is an answer. Before the Original Light was Dark Light. I did not say there was nothing. I did not say there was an eternal void. There was Dark Light. And, as the untold time of existence passes, the presence of the force of Dark Light as a part of existence must be recognized and acknowledged. It is for this purpose that I am here to share with you.

You are here in the service of the Universe and, in the preparation for the totality of your service; each person is in a process of purification and readjustment. All patterns of expression that are not within the expressive truth of soul are gradually dissipating and leaving your presence. What does that mean in simpler terms to you? The Dark Light is leaving so the Light can enter the Original Light of your souls.

All humanity must begin to understand that, in the pursuit of conscious evolution, when one searches and reaches for oneness with soul the path is Dark. The experiences are ones of the absence of Light for without the absence of Light you could not recognize the Light.

I urge you most sincerely to review in your mind the pages of your book of life. Recognize the Dark Light of actions others have taken against you and then the door will open for you. It then becomes your choice. You may invite the Light to enter. If a situation comes into your consciousness, where there was no experience of Dark Light, be aware that it will appear in your life.

Without the experience of Dark Light there will never be trust; there will never be a knowing within you. There will always be doubts in your mind relating to what you consider to be your truth.

The Law of your planet Earth, the foundation of your karmic experience of life, is based upon the experience of choice, of the polarity of energies that are involved within the truth expression of your soul. The shifting of the truth expression on this polarized frequency of Light and Dark is in constant motion. It cannot reside at peace for as long as men and women walk on Earth in the grip of the energies of emotions. The determination of truth must be a variable energy.

When an experience occurs that is in a shade of Gray, the truth is received in a Gray Light. The closer the experience is to the full expression of the Dark Light, the closer comes the expression of the fullness of the Light. At times you have referred to this as removing the layers of the onion one by one. A most satisfactory description. It is not that you must endure a state of chaos but you must endure the void of Dark Light, the surrender. In the surrender, all shades of Gray are gone and the purity of Dark Light descends upon you and you rest in the void to surrender the old patterns. Soon, the Light begins to rise over the horizon; resistance is gone for you have surrendered to the truth of soul.

This is the journey you travel at this time. It does not of necessity generate pain. Certainly it generates emptiness. Certainly it destroys your foundations of the past but these foundations were constantly shifting for you and, at times, you drifted on the waters. The void of Dark Light is indeed the first step up the mountain. Rejoice upon the apex and bathe in the Light of your soul.

This, beloved children, is the destination of the journey you travel now. Everything will be done from all dimensions of existence to support your decision to Light the candle of your soul.

The Aura

The aura is a vibrant field of energy that emanates from the Third Eye chakra of the physical body. We call this chakra the seal of the aura, as it protects the body from invasion by outside energies and expressions of negativity.

The aura is composed from three different sources of energy. They are the radiant energy of your soul, the magnetic meridian of your body, and the Earth meridians of your emotions and life expressive energies.

The auric field of energy extends to a distance of thirteen feet or three meters away from the body. It is a pulsating, constantly changing radiating light. It reflects our mental, physical, emotional, and spiritual condition of life at any given moment.

For Example

1. When we are angry or upset, our aura becomes "hot," and presents the color frequency of energy that seems red, and emits quite a warm sensation or feeling. Our perception of this is becoming flushed and red-faced, sweating, or just feeling warm all over.
2. When we are involved in a spiritual expression, such as meditation or in a house of worship, our aura becomes mainly a soft blue combined with white. This reflects our awareness of and presence with God. The colors and energy are calm, serene, and many times just make us feel better about ourselves.
3. If you are healing someone, your aura will begin to project itself in various shades of green and pale rose, depending on the type of healing being performed. The color green is referred to as the color of healing of all conditions. The rose relates to Universal health of soul.
4. If you are ill, tired, or weak from stress, your aura will reflect this in the dullness of the color. It will lose its vibrancy and clarity of tone. This always signifies a weakened interior energy condition.

Every disorder inside the body has an energy reaction that becomes transferred to the auric field around the body. If you have had surgery twenty years ago, there will be scar energies detectable in the auric field. Any expression of disorder inside the body will project as a hot spot in the auric field. These can be detected by passing the hand slowly over the body approximately four inches, or ten centimeters above the surface of the skin. By doing this, if there is any area that contains a form of infection or disease, you will feel a slight change in the heat of the aura over that area of the body.

The aura can be utilized as a medium for healing. We have said that if a person does not believe in the existence of God, you cannot heal them with God's Light. However, everyone is still entitled to a healing, as God is Divine Grace. Under these circumstances, you may apply what is called, auric healing.

This form of healing is most effective when the person is standing erect. In this position, the entire body and the aura are straight and flowing evenly out of the Third Eye, and down the length of the body. The procedure is as follows.

You are standing next to the person, and as you breathe, begin to draw energy into your Crown chakra. From your mind send the energies down to your hands. In a few minutes you will begin to feel tingling sensations in the energy insertion points on your fingertips. Close your eyes and be aware of the energy presence on the surface of your hands and fingers. We normally feel things through physical touch pressure on an object. Energy is different. We need to develop the ability to sense energy through vibrations and temperature. When you are aware of the feeling sensations, hold your hands four inches, or nine centimeters over the top of the person's head. Very slowly begin to move them down, one hand on each side of the body following its shape as your hands descend. Try to be aware of what you feel as you move your hands. If you reach an area where you feel a change in the temperature or heat of the energy, pause and try to locate the exact place that is the source of the change. When you have reached their feet, touch your hands on the floor and mentally transmute the energy by returning it to the Earth.

Now you move, and repeat the process over the front and back and on the sides. There are four positions around the body where you repeat the process. After each position return the energies to the Earth for transmutation.

This is known as aura cleansing. The purpose is to remove any energy dirt that may have accumulated in the aura during the day, or number of days. We all attract a measure of negativity from other people. This will remain in the energy of the aura until it is cleansed.

It is also effective to cleanse the aura when the person is lying down. The process will be to sweep one side of the body at a time. The person will need to turn from one position to another, until all four sides are completed.

There is a significant amount of research and experimentation being conducted on the energy of the aura. In time, it will be used as a completely accurate diagnostic tool for isolating and locating area of disease.

Cellular Memory Healing

When a soul enters a fetus at the moment of conception, its energy completes the trinity of creation — the egg, the sperm, and the soul. Sometimes we refer to this expression as being created in the image of God, as our soul is composed of male, female and love energies.

As the fertilized egg begins to divide and multiply, the soul becomes an integral part of the process and adds its essence of existence to each additional cell. A child is created — the ultimate miracle — and in every cell of its being is the essence and the memory of the soul.

To present a simple definition of cellular memory is not a complicated task. To accept the definition requires you to have an open mind, and to consider that all possibilities exist in all circumstances of life. If you are able to think along these lines, a world of new concepts and realities will open for you.

Cellular memory is the energy presence of the sum total of every experience the soul has completed during the total span of its existence. This, basically, pertains for our purposes to this current lifetime, but involves the balance of the soul's expressions on Earth as well as other dimensional expressions both physical and as pure energy.

If the following statements can be part of your spiritual belief system, you may begin to understand the orchestration of life here on Earth.

- The existence of the soul is eternal.
- The soul incarnates in increasingly more complex civilizations to advance its experiences, learning, and evolution.
- Energy is eternal. Every thought and action generates a reactive energy pattern that eventually becomes part of the soul's evolution.

These statements tell us that the reactive energy of every experience you complete — or have imposed on you by someone else — remains in the cells of your body for your entire lifetime.

These reactive energies become the foundation of the automatic response patterns in your subconscious mind. Often we refer to this as the karmic lessons and experiences of life.

When the current lifetime has been completed, the soul absorbs all the energy of the personality's experiences and the unconscious patterns of reaction. This, in effect, becomes the soul's growth for the incarnation and is present eternally in the memory bank of the soul.

In every society on Earth, the majority of automatic response patterns are created from the result of emotional experiences and free-will choice patterns of decision that lead to actions. As a child, these are usually imposed on you by your family and your social environment. As the reactive energy is created, it settles into the body's cells that are associated with the original action. The energy remains there and responds from that area of the body memory in the future.

The Universe tells us that all energy expressions are in ascending and descending polarities.

- A soul is created; a child is born.
- A soul learns that other souls exist; a child discovers other people.
- A soul begins to relate to other souls; a child relates to other children.
- A soul begins to evolve with exposure to higher energies; the child begins to learn in school.
- Both soul and child eventually become involved in higher education.

So here we are in a life on our planet Earth! Why Earth? After all, astronomers tell us there are more than 300,000,000 planets in our galaxy alone.

Why do we take mathematics in school? Because it is a part of the whole, a part of the completion of our education. The soul comes to planet Earth for specific classes of experience to advance its education.

The unique aspects of life on Earth involve the presence of an emotional pattern of expression and reaction, as well as the most important factor of our lives, our free-will choice.

Let us consider a concept. When a soul elects a lifetime on Earth, it decides what classes it wishes to experience and complete as part of its evolution. It chooses the parents according to their genetic and hereditary factors, as well as their conscious personalities. It chooses the astro-astrological timing of birth. It chooses the vibrations of a name as well. It does whatever it can to ensure that the exposure in life will result in completing the experiences it has chosen for that life.

Do you know what that means? It means that you and everyone else are perfect for your soul! You are exactly what your soul wanted you to be.

You are not:
- Too short or too tall.
- Too fat or too thin.
- Your breasts are not too large or too small.
- The features of your face are perfect for you.
- You are supposed to be a woman/man.

Can you finally accept yourself? **YOU ARE PERFECT FOR YOUR SOUL!**

I have listened to the stories of many people about their conditioning from their past-life experiences. They call them past life karma. This may be accurate if their definition of karma is experience, not punishment.

As life progresses and maturity begins to express itself, the memories in the cells of the body awaken. As this takes place, our emotions, our actions and feelings can be affected by the cellular memory of the past-life experiences of our soul. The result of past life conditioning can cause us to exhibit some patterns of abnormal behavior such as fear of water, height, animals, confrontation, and many other seemingly unexplainable behavior reactions in life.

The percentage of automatic responses remaining from past life experiences is very small. The majority of these patterns are the result of the conditioning and dysfunctional expressions of the current lifetime. These conditions need to be brought to the surface and dealt with, if the person is to become free of the undesirable reactive expressions.

We frequently call these patterns the shadows of the soul. By definition, a shadow is an "un"-experienced pattern of energy in the unconscious memory centers. The shadow will always exist, as energy is eternal. With the proper understanding and treatment, they can be made passive expressions of memory from which the person can learn and consciously evolve.

The most important reactive shadow cell memory centers are:

The Second Chakra

This spiritual energy center contains all the cellular memory resulting from dysfunctional behavior and rejection. It records and stores all emotional reactive energies that result from the experiences in life. The resulting effect is to generate a subconscious program that affects all future behavior, both positive and negative.

a) The experience: Not receiving enough love as a child.
The program: I am not worthy of love.

b) The experience: Physical, mental, emotional or sexual abuse.
The program: It was my fault. I was bad.

c) The experience: Parental separation or divorce.
The program: People I love abandon me.

d) The experience: Being raised in sexual fear or shame.
The program: My body is dirty and evil.

e) The experience: Being born male/female and parents wanted the opposite.
The program: I am an unwanted child.

f) The experience: Marital infidelity.
The program: I am a failure as a husband or wife.

As the results of the spiritual counseling and healing of many people, I have determined that the patterns listed above are the most common reactive expressions stored in the cellular memory of the Second chakra. The average person walks through life completely unaware of the programming. All they know is that something is wrong and they are not able to feel fulfilled in their life expression.

The Solar Plexus Chakra

This chakra retains all the cell memory of any actions that resulted in sacrifice. The repetition of these actions eventually generates the expression of emotional guilt in relation to self or to others.

a) The experience: Telling people what you think they want to hear.
The program: If I please them, they will accept me.

b) The experience: Always making everyone happy.
The program: I avoid confrontation, and always keep peace at any price.

c) The experience: Taking actions that are not in truth.
The program: If I do what they want, maybe they will finally love me.

The majority of the actions that affect the Solar Plexus chakra have been taken as the result of the programming in the Second chakra. Whenever we feel that something is wrong or lacking in our lives, we take different actions to discover and correct that situation, even at the subconscious level.

The most common actions of sacrifice are:
- Seeking approval from others.
- Justifying negative parental actions.
- Seeking life-long security.
- Looking for someone to say, "I love you."
- Always keeping peace. Don't rock the boat.
- Playing roles to avoid discovery.
- Sabotaging yourself.
- Doing anything to avoid confrontation.

- Always being just what others want you to be.

The best way to begin to move yourself out of actions of sacrifice is to pause before making any decision or taking any action. Take a breath, and ask yourself one question. "If I do this, will I be in sacrifice, or is it really my truth?" Keeping yourself aware of how easily you fall into the trap will begin you on the path of reprogramming the existing cell memory of your actions.

The Right Breast

This unique energy center contains the energies of self-love in both men and women. The absence of self-love comes from not accepting your physical structure and its proportions. We can call this the absence of self-Grace by not accepting yourself as you are today. This condition can also result from repeated rejection patterns that have created a negative attitude towards yourself.

The Left Breast

The tissue in the left breast contains the energy of prolonged anger, resentment, and judgment against yourself or the actions of others. Many times this condition manifests when we point a finger of accusation at someone else. We must begin to accept the total responsibility for our own actions and life.

The discovery of shadow energies and unwanted cellular memory patterns can be realized through proper counseling techniques and the sensing of energy variations in the designated area of the body. Most of these conditions can be relieved through the use of replacement experiences. By offering a person a choice, they begin to realize that they can alter their behavior and lead a more fruitful life.

With the use of meditation and guided imagery techniques, we can create the experience of the polarity of the existing shadow. All energy patterns of expression exist in pure polarity inside of us. This is the foundation of the free-will choice expression of life. The most common ones are:

Love	Hate
Success	Failure
Worthiness	Unworthiness
Judgment	Grace
Rejection	Acceptance
Abandonment	Unity
Sacrifice	Truth
Joy	Sadness
Peace	Turmoil

If you tell me that you always feel unworthy, I will say to you, "Come and let us look inside. There must be a pattern of worthiness. Let's find it. We can. It has to be there. It is the law of energy polarities."

You can apply this concept to anything, any pattern you consciously express that is not satisfactory for you. It all comes from the power of your mind. This is the pure expression of free-will choice. There is always a choice in everything in life - without exception.

If you have an initial experience of a desired polarity expression, by repeating it on a daily basis, the body will accept the new truth program within thirty days. This will de-activate the old shadow pattern and place it in a passive, memory position for future reference and learned wisdom of choice.

You are the conscious expression of your soul. That means that it is your responsibility to create the actions and inner patterns of response that are in accordance with the desire and will of your soul. No one is here to be punished for anything, or any past pattern of cell memory. Those are illusions.

We are all, and I mean all, here to have a life of personal freedom, peace, and joy in our lives. We are in control, if we only make the efforts necessary to achieve this condition of life.

Success and Failure Mechanisms

The polarity expression of success and failure is among the main patterns that are responsible for many of the causes of the expression of disease, as well as the creation of mind-judgments that generate the spiritual program for the disease.

In order to successfully apply this mechanism, you must understand the divisions that are responsible, and participate in this program. We have associated an aspect of success with each of the letters in the word to correlate an association for you.

1. Sense of direction and a goal.

Those who believe they can, do. Those who doubt, stay at home. In order for you to achieve any degree of success, you must have a direction. You need to know where you are going. There must be something to be successful with; and so we are going to deal with goals.

A goal is established from what we have determined to be a valid want or need. If we establish a goal from a desire, it will not be valid; as desires are purely emotional and often illusions of ego. Something that is a want starts out as a desire, but you can change the perspective by acknowledging that it is not a need but you want it. After all, we are entitled to have things we want as well as the things we need. Most of our goals will come from our needs. We need to determine the validity of a prospective need.

In setting goals, you must be aware that there are two categories of goals, short-term goals and long-term goals. The average person does not establish short-term goals. They look at the overall picture, at the end result that they want to achieve, and set that as the goal for themselves. It is for this reason that so many people end up in what they call failure.
We cannot ignore the positive effects of psychology. They remind us that the best incentive in life is a successful experience.
For example: if you set a goal for yourself to acquire $10,000.00, that is great but you are not taking it far enough.

You need to establish monthly goals of saving money. If you decide to save $300.00 a month, that is what your short-term goal should be. If you do not have the interim goals, it will take you too long to experience the success. The saving of the $10,000.00 will seem too far away and you could find your motivation dissolving in three or four months. If you save the money monthly, you will feel the success each month and have a new stimulus towards reaching your long-range goal.

This psychological tool is most important for everyone to use. It sets you into achieving a pattern of success. Small successes — one after another — create a success orientation.

Goals are not always constant. There are many times when a re-evaluation takes place and the goals are expanded. If you had set your goal to save the $10,000.00, by the time you had successfully saved $6,000.00, you might decide to reset your goal to $15,000.00. This would be a very positive action. As the result of your interim successes, your motivation has become stronger and your success orientation will stimulate the further successes.

The difference between success and failure is an intangible figure. You might feel very successful earning $20,000.00 a year. You may have a friend who earns twice that much, and quietly considers you to be a failure. Success is always relative to the goals the individual has established for themselves. Live your life with the glass of water half full, never half empty.

There is no concrete definition or description that can set guidelines to label the results of any action as failure.
If you can accept the polarity of this energy, you live and act with the varying degrees of greater and lesser success.

When you are trying to establish your goals and determine a direction and course of action, take time, sit down and meditate, and validate the direction for yourself. Is this valid for you or is it something they want you to do? If you cannot truthfully validate the goal for yourself, you will not reach the higher level of success you desire.

Success is an emotional response; and even if you reach the goal, you would not consider yourself successful if the selection of the goal was taken under pressure from other people, or chosen to please and make them happy. For example: your parents want you to attend university. You don't like school but go to please them. Even if you graduate, you will not feel successful. You will feel sacrificial and unsatisfied.

I repeat: success is a personal emotional experience relating to an action that you take in your truth and your choice. If you take the action to please others, you will never feel successful. The reality is that you will have made yourself a victim to please other people.

2. Understanding

We need to remember that it does not matter who is right or wrong in any given situation. It only matters what is right. This correlates with the words: it is always better to understand a little than to misunderstand a lot.
How does this word "understanding" apply to your expression of life? It has a dual meaning, relating to understanding others as well as yourself. Do you understand yourself? If you do not, how can you ever expect to understand someone else? You will always see them in your own personal mirror of your self-considered image.

Understanding yourself is composed of many facets of expression.

- Are you gregarious and outgoing?
- Are you shy and retiring?
- What about life is really important to you?
- What do you want to achieve in life?
- Are you happy and secure with a partner and children sharing your life?
- Are you the type of person that needs to be out in the world achieving your personal goals all the time?
- Are you the type of person who can sit in an office all day long?
- Do you need to interchange with people in your work situation?
- How much love do you need?
- How much love are you willing to share with others?

Building Confidence

When I say to you that your existence is the power of your mind, I am trying to create an understanding inside of you. I want you to try and accept that every emotional reaction and pattern you have is subject to change. It is never too late in life to change. Years have no relationship to change — only your attitude. Do you allow your mind to relax and sit in the rocking chair, or can you accept the mind's eternal power and ability to continually expand?

All our emotional patterns of expression are connected and in conjunction with each other. One fallacy directly affects the positive expression of other patterns. We have spoken to you of the success/failure syndrome. Those expressions are deeply affected by the limitations of our self-confidence.

When we experience a lack of self-confidence, it is not the result of the difficulties that we have encountered. The difficulties are created by the lack of confidence. To assist you, we are going to discuss twelve steps that can be utilized in the building of self-confidence.

1 When a soul comes to incarnate on Earth, it comes here for a lifetime of happiness and joy that is created from growth and evolution. One could say that if this is true, then our main goal is to live and be happy! That also means that happiness is a state of confidence in oneself — a great goal for you to have. If we can accept the supposition that confidence is this condition, then we must, of necessity, create a definition of happiness for ourselves.

Do you like yourself?

Are you satisfied with yourself when you look in the mirror?

Are you goal-oriented and achieving?

Are you the best you can be?

If you can answer, yes, to these questions, you certainly have a good degree of self-confidence. However, don't become complacent, remember the Law of Eternal Change. Whenever you sit back and say, "I've got it made", you are left behind, as the world continues to progress and evolve.

2. You cannot willfully force self-confidence on yourself. It would be nice if you could just sit down and say, "I am very confident about me" and have it. First experience who you are. What are your shortcomings? In what areas of life are you excellent?

When you set a goal for yourself in an area that you do not excel in, you should re-examine that goal. It is like the person who wants to be a surgeon, but has arthritis in his hands, or a person who wants to study nuclear physics but has trouble with simple mathematics. Examine the proposed project, and look at your confidence level in relation to your capabilities to fulfill and reach the goal. If you are not properly equipped, re-schedule and re-adjust the goal. We rarely take the time to become involved in this form of analysis. How many times have you looked at what you are capable of achieving successfully so you can have the proper confidence in yourself?

In a spiritual sense, the knowledge that you are a child of the Universe and have your personal relationship with your Creator can be one of the greatest confidence builders for you. It creates the knowing and understanding that you do not have to be an expert in all things. We each have our special soul, with its unique talents and abilities.

Many times people come to me and ask me what area should they express for their service. What was the purpose for their soul coming into this lifetime? My answer is always the same.

A soul does not normally incarnate for a specific expression of service. They are just to serve with their energies. The conscious expression of service is the choice of the personality. So, you are free to do what brings you the joy of life!

What area are you drawn to? Where is your comfort and your confidence? How will you serve people in the most complete way?

If I were to say to you that you were here to be an astrologer, what if you don't like astrology? I am not going to say, "Too bad. Go study astrology."

We cannot perform to the best of our abilities, if we do not have full confidence in them to achieve our goals successfully. That is the bottom line of all decisions and actions of life.

3. Confidence means positive thinking and positive actions. You must have the goal, and so you must imagine that you are already at the goal. We call this process visualization.

The salesperson goes to call on a customer. He does not have too much confidence in himself, so he sits in his car, closes his eyes and creates a play in his mind. He visualizes himself entering the house, making his sales presentation, and the client smiling and signing the order. Now he allows himself to feel the emotional success of making a sale, before he has even seen the client. When he opens his eyes, he has already experienced the emotions of success. His attitude and the confidence in his abilities to succeed will work; and his business will become much more successful.

We can apply this process of self-imagery to every aspect of our lives. It is a widely accepted and used tool by many athletes to enhance their performances. The main key to this process of visualization is to have the emotional experience of success at the end of the process. This begins to program the building of confidence and emotional success into your life.

4. Everyone becomes involved in actions that result in varying degrees of the energies of success. This is the place we must condition our mind to relate and react, not in degrees of failure. Whenever you are facing a new situation that you feel might be difficult for you, remember your past successes and activate the good emotional feelings you had. Use those emotions to strengthen you expression of confidence.

No one can say: "I don't have any successes in my life. Everything has always been a failure." That cannot be valid. Success is a variable experience that touches everyone each day of their lives.

If we concentrate on our successes and not on our failures, our attitudes will always be positive, and the old cliché, to have your glass of water half full, not half empty will apply in all your life situations. Nothing succeeds like success, and nothing builds confidence as well as the recalling and experiencing of your conditioned responses to your past successes.

5. If you have the desire, a wonderful feeling of enthusiasm will lead you into the knowing that you can reach your goal. It is for this reason that I have suggested to establish shorter-term goals. In this way, the desire, the emotional response and the adrenaline flowing throughout you will act as a stimulus, creating the confidence for you to reach that goal.

It is very difficult to be involved in something that you do not totally believe in. There are many people on their spiritual path who find themselves in a situation that was unexpected for them. All of a sudden they are having difficulty dealing with their work situation. They find it difficult to remain in what we call the business energies. These are the little white lies, the pressure to sell, sell, and sell, even if they don't believe in the product.

What are they supposed to do? The only answer I have for them is to quit the job. Why? If a person comes to a place of understanding that the belief in what they are offering is no longer valid for them, they have no goals to reach. Their success will begin to diminish as their motivation dissipates. They have lost their desire to promote their product, and this attitude will be reflected to the customers and have a negative result on their success.

If you cannot be enthusiastic about a goal, and have what we call a good gut feeling and believe in it, walk away. It will not be a valid goal for you, and the first time you are faced with an objection, you will collapse. I have seen many people lecturing. Once in a while, when it is time to answer some questions, the lecturer becomes flustered and insecure. The audience senses this, and may begin to wonder if they really believed what they had spoken about. In the reality of the situation, we have to say that they probably did not have enough confidence in themselves to answer questions that were not part of their prepared notes or material.

One question made a crack in their wall. Their belief had no foundation of confidence in their own knowledge.

6. Try to acquire the habit of confidence. Confidence is built on past experience or, as I said before, nothing succeeds like success. Every time you sit down and try to plan a new project for yourself, remember the results of former projects that you planned, fulfilled, and were successful. Build your reservoir of past responses and be aware of the role your subconscious mind plays for you. It is the storage center for the actions and reactions of all past situations in your life.

When someone is faced with situations that they have not been involved with in the past, they can have insecure reactions. They may get a fever, a migraine headache, or their stomach will react like a volcano. This happens to certain people every time they are threatened by having to deal with a new situation. The perfect example would be a test in school.

We have all said, "I know my work, but for some reason, every time I sit down to take a test, my mind becomes a blank and I get butterflies in my stomach." This is the lack of confidence in successfully dealing with the unknown.

If you were to sit down, close your eyes and review the material of the test in your mind before the test, you could acknowledge to yourself that you have assimilated the material. When the time comes to take the test, you will smile with all the confidence you need. Why? Because you already experienced the material and you have nothing to be concerned about.

7. Activate your subconscious and re-experience the confidence you have expressed in the past. Pull all of your successes out of the files and involve yourself in the thoughts and memories that are positive for you. All the negative experiences of the past are merely tools for you to learn from for the future. Once again, there is no failure.

When we activate the subconscious mind to release the successful past experiences into consciousness and our emotions, we never have to be concerned about having confidence.

It will always be there, and the new attitude will say, "It doesn't matter what I try, I know that I am the kind of person who can do, whatever I believe I can do. I have a lot of confidence in my abilities to succeed." Isn't that better than standing there shaking all over, wondering if you can do something? Instead of saying that you don't know if you can do that, ask yourself how successful can you be in that new experience.

You are creating the positive conditioning. By doing this, you will determine the degree of proficiency and success you are able to achieve in that area of expression. You will determine your qualifications, and from the results of past involvement, remember what not to do in your future efforts. If you become a victim to a lack of self-confidence, you will create a program of avoidance and mentally generate the reasons not to become involved in the situation.

8.   If you have had a program telling you that you have never had a successful experience, imagine in your mind how you would feel with the experience of success. Make a playhouse in your mind, and enjoy the wonderful feeling. The subconscious mind cannot tell the difference between a true experience, and one that you just vividly create from your imagination. This tool is most successful in areas where people feel inadequate in these situations.

- The full expression of their emotions.
- Their relationship to money.
- Accepting supervisory positions at work.
- Their abilities to heal, counsel, and express their spiritual truth.

The major key is to generate the success prior to the experience taking place. This eliminates the feeling of dealing with something new and unexperienced. The situation will seem like an old friend, not as the enemy.

9.   If you have a need to worry, make it a constructive project in relation to a future goal. Work slowly and progressively towards the goal. If you still need to worry, think about how you will handle the success, not the failure. I have met many people that seem to have the need to worry all the time. If they do not have a problem, they create one. I remember when my father came down to breakfast he always said that he didn't sleep a wink the whole night. I used to wonder as a child.

How that was possible? Eventually, everyone just ignored those words. We all learned that it was a way to attract attention and sympathy.

This attitude is not always negative. Worry does create some stress, and stress is an incentive when used properly. It starts the juices flowing, and becomes a stimulus to get moving, begin doing, and taking the actions that takes the worry out of your emotions. Worry without action creates depression and eventually may cause someone to withdraw from life. Worry becomes concern, then doubt and eventually judgment. If you create these self-judgments, you will place yourself into avoidance and begin to avoid all situations of a positive nature. Worry does not necessarily mean concern, but concern will lead to doubt and judgment of yourself.

When you have a disagreement with someone and worry about losing your friendship, do you go to the person and say, "I think we may have a problem that we need to discuss and resolve? I don't want it to cause a problem in our relationship." In this case, the worry becomes a stimulus for taking the action of healing and positive resolution.

Why do I worry so much all the time?

Do I worry because I don't have confidence in myself?

Do I worry because I am a procrastinator?

Why can't I make decisions? What am I afraid of?

Indecision becomes the incentive to sit, eat, watch television and perpetuate postponement that creates increasing pressure and tension in life.

10. If you will accept negative feelings as a challenge for yourself, you will remember that confidence gives you the power to rise above all negative feelings, even failure. It is possible to take failure and use it as a challenge for the creation of success.

Here we have another polarity of energy that is expressed and controlled by the power of the mind. I am speaking of positive and negative energy. It is always a good policy to look inside of words.

Is there really a universal energy of negativity? How could that be possible? The soul of the Creator-God is the essence of Divine Love and Grace. In the purest sense, that energy is all that exists in the universe, nothing else.

Here we are on Earth, and we have negative energy. Can you consider the possibility that we create the negative energy we need from Light, in order to have our karmic experiences of choice? How else would it be possible to make mistakes? We could never grow if everything was only one truth.

If you can accept this belief, then everything can change for you. If you do, then you can have the control, and can choose and empower energy with your mind to fill your needs and reach your goals. Through the power of your mind, you can create the realities of your life!

What is it that makes one person an Olympic champion and another who always comes in second? The champion uses the possibility of failure as a challenge or obstacle to overcome. They know they can do better and can reach the achievement of their goals. They take the so-called negativity and polarize it into the incentive power of success.

The person, who experiences the negative result of an action and has a pity party for themselves, creates the confirmation of their lack of confidence in their abilities to succeed. Learn to love a challenge and to be an explorer. Let a challenge be exciting for you. If you include the results of your past successes as part of your new challenge, your confidence will soar into the skies and you will achieve future success.

If you feel comfortable about a situation, accept the challenge. Allow yourself healthy worry and concern. It will be the stimulus for you to sit down, go through the process of either accepting the challenge, or searching for an alternative. A major problem can raise its ugly head. In making your decision, are you going to compare yourself to someone else? If you do, you are doomed to failure.

We are all unique. There are no two souls exactly alike, and certainly no two people with identical conscious and unconscious personalities.

I believe that, in everyone's uniqueness, we all have one ability that stands out like a shining star in our lives. Go inside, discover what it is, and use it to bring the great joy and achievement into your life.

11. Try to substitute a good feeling of confidence for a bad feeling of frustration. Make it a habit, and remember your past confident expressions. This is what we call the Universal Law of Allowance.

The Law states: It is in order to accept the existence of all things and people as they appear to you, without judgment. It sounds quite simple, and it is. It does not mean that you must accept everyone in your life, or every action taken by other people as your truth. It is always your right to make your own decisions, but not to judge the decisions and actions of others.

The compliance with this Law eliminates actions of comparison, judgment, and expectations. Everyone becomes free, and responsible only for themselves. If you want to be lazy and do nothing, enjoy it. If you do not want to make your bed in the morning, don't. It is your bed. Can you imagine that taking these actions will build your self-confidence? It will! You are in the "being" of responsibility for yourself. You do not have to defend your actions or lack of actions. You are in charge of you, and all the choices you make in your life.

I know those words sound like a wonderful dream; and they are. The danger develops when being good to yourself for a while becomes a pattern of avoiding life. We can very easily begin to walk on the treadmill of nothingness: I have this big problem; I think I will take a nap. Or all of a sudden I feel tired, like I have no energy. What a great way to avoid life, and walk into the world of nothingness.

12. Our final rule applies to something we do to ourselves repeatedly. We listen to the old tapes of frustration and unhappiness in our mind. Now we need to create a new program of confidence and happiness. How? Believe in yourself in an active expression, not a passive wish of illusion.

Many spiritual people bury themselves in old karmic situations or past life experiences of their soul.

They adopt conditions that the former personalities of their soul have experienced with negative results from the past lifetimes. They walk around saying, "I am here to be punished in this life." They are playing the old tapes of unhappiness and failure. These become their success and confirmation of their unworthiness in life.

Most people have been exposed to a form of dysfunctional behavior in their childhood. It was either in the form of physical, mental or emotional abuse. Some people keep playing the tapes of unhappiness and negativity. They are unable to release them, or they need to keep the suffering active and as a tool for their failure in life. The old tapes keep reinforcing the lack of confidence.

Once again, I must mention the Laws of Allowance. You need to acknowledge what happened to you. You need to acknowledge past unsuccessful actions. They all happened and were real events of your life. What happened to you?

You were abused.

You were ignored.

They beat you.

They raped you.

These things happened! Acknowledge in your conscious mind that they happened. This is the only way you can take away the energy of the power and control they have over your life. You cannot replace the action, but you can replace the reaction, by reprogramming from your power of mind.

In the allowing and accepting of all things that have no role in your life now, and for the future, you can positively create a new mind consciousness. "Because I have acknowledged that these things occurred, and have allowed them to occur, I can let go of judgments and anger. I can say that I no longer have the need for these things in my life. I replace all old tapes with their polarities that will fulfill my life with happiness, confidence, and success."

Years ago, a young couple came to see me who were both verbally abused by their parents. The abuse was repeatedly reinforced about their unworthiness and how useless they would be in life. They met each other, discovered their common pattern of life, and decided that they could help each other in life.

She accidentally became pregnant, for they had decided never to have children because of what happened to the both of them. As a result, the child grew up with many problems. She never experienced discipline or direction in life. They could not even say "no" to the child, as they always remembered what happened to them.

When they occasionally vented anger at the child out of frustration, they began to cry, and hugged, and apologized to the child. Their own abuse came alive, and they were being abused once again. The child began to control the parents and use them against each other for her own gain. They could not resist and overcome their old tapes.

If you desire to utilize your spiritual growth as a part of your future, it is very important for you to be a confident human being. This confidence needs to relate to you as a Child of Light, a person who accepts and experiences God within themselves. When you allow the fire of your soul to begin to warm your heart, you will have no fears.

In the course of your growth, you will endure testing situations designed to present you with obstacles for you to overcome. This will assist you in determining your truth in many avenues of expression. Your path of growth will never be totally smooth, as the discerning process never ends. It is a timeless process of evolution.

Each time you are faced with a new situation, you must take a risk, and if your do not have enough confidence in yourself, you will back away from your path. For every ten people who begin a path of spiritual growth, half of them fall away when they reach the point of self-examination of their confidence in themselves. If the confidence is not there, fear takes over and all the old thoughts return to their mind.

If you want to be a healer you cannot stand with a client and hope that it works. You need to remember your successes.

You need to have the belief and faith along with the confidence in your abilities to transmit energy to someone. If you play the old tapes of frustration and failure nothing will be accomplished.

The vibrations of emotions and spiritual energies are polarities that do not mix or integrate with each other. When you doubt yourself, you are in emotional distress and cannot channel healing energies through your body. Having confidence in yourself as that Child of Light, brings peace into your heart and into your expression as a healer.
Total confidence cannot be achieved in one day. It takes time. You need to build on your successes and the positive results of your efforts.

Over the years, I have been involved in many hundreds of healings. Not everyone has resulted in the curing process. That does not mean that I will generate doubts in my mind in relation to my validity as a healer. I always remember the people who thank me after a healing with a smile on their face and a light in their eyes.
I do not forget those who remain with disease, but I understand that as long as I do my best that is all that is asked of me. I can acknowledge the following to myself, "I may not be the best healer in the world. Not everyone who comes to me will become cured, but I need to keep touching and healing; for those who do respond smile from their hearts."

The universe has given us a multitude of tools to use. Think of them. Be aware of what they are. Your life is now and in your anticipated future. The old tapes and old actions are part of the past. We can use the knowledge of the old tapes to build confidence for the future, not to be a chain around our neck that drags us down.

Look at everything and sift through it, and sort it out. Look over what is not positive for you, experience it and grow from the results it created for you. From the residue of the negative or bad experiences, you can create the confidence for now and tomorrow. Now is the time to do it all.
The achieving of your goal to become a confident person is also closely interrelated with another polarity of energies — the expressions of happiness and unhappiness. I do not have definitions for you, as there are always variables depending on the expression from your mind's past history of your life. I have assimilated examples of both for you. Where do you live, and where do you wish to move?

Happiness

Happiness is the search to self-fulfillment.

Happiness is a state of mind having pleasant thoughts most of the time.

Happiness is a goal in itself. A state of being.

Happiness means reaching for goals with courage and understanding.

Happiness means you are always true to yourself.

Happiness means that you have compassion and live through your mistakes.

Happiness means never retiring from the expansion of life.

Happiness means spiritual and emotional freedom.

Happiness is a contagious disease.

Unhappiness

Unhappiness creates loneliness.

Unhappiness means the loss of your true identity.

Unhappiness creates limitations and separation from self.

Unhappiness means over-reacting to stress.

Unhappiness means accepting negative thoughts in your mind.

Unhappiness means never confronting lies.

Unhappiness means always being the good person.

No one can make you unhappy without your consent. If you realize this, you have the ability to turn any negative into the polarized positive. Happiness is a very contagious disease. Expose yourself to it with confidence.

We stepped into a new millennium. Now we cannot leave our responsibility for our actions, thoughts, and decisions to others. In the age of Pisces, the Masters and Gurus have accomplished their function. We walked into the age of self-mastery and personal freedom. Who, besides you, knows your true realities? None. Others see you in their own reflection and have expectations of you. Through this, they create their own securities to you, and this is not always in your best interest.

You are the expression of your mind. The more you realize and the more you live your life with this attitude, the greater are your inner and outer benefits.
I am the conscious expression of my soul! You are the conscious expression of your soul! – Your friend, your partner, your teacher...

Your connection with the Creator God! Be who you are, and not what somebody else would like you to be.

Live a Life of Joy!

The 36 Atlantean Chants

Chant 1: A-LEO-U
This chant stimulates the energies of the body meridians. An activation of unconscious memory patterns take place and move them from dominant to active expressions of knowledge.

Chant 2: BA-EO
The chant creates a resonance that activates related brain centers to remove distorted energy flows between emotional and soul energies.

Chant 3: GA-MO-AL
This resonance can have a positive energy effect on conditions in the liver, pancreas, spleen and kidney organs. It can be used with the 17th ray, persimmon.

Chant 4: DO-LAE
This chant is used to ease conditions of emotional stress relating to social and physical life.

Chant 5: HA-JO-HA
These are the vibrations to call upon God's light to heal with the power of regeneration. It is associated with the 22nd ray, pearlized white. Invoke this resonance during the healing process.

Chant 6: VO-A-A
The chant brings your deepest vibrations into consciousness and begins the journey into your soul energy.

Chant 7: ZOO-UR
This resonance begins to loosen blocked energy to begin the release of unwanted karmic patterns of expressions.

Chant 8: CHO-RA

The chant creates the courage to express positive actions and counter-balance previous negative actions and judgments.

Chant 9: TU-LA-RO
Use this chant with conditions of arthritis, nerve conditions and skin eruptions. It is associated with the color pale rose.

Chant 10: YO-OOH-DA

These are the sounds of the God within. These are chanted in conjunction with the 36th ray; the perception of color. Use it for spiritual self-healing and refinement.

Chant 11 CO-LAE-AH

This applies to the healing of physical deformities, broken bones, after they have been reset, healing after surgery etc. The chant is used in association with the color red. This is mainly a physical Earth vibration.

Chant 12 LAU-RR-U

This chant symbolizes the energy of development and evolution. It enhances the ability to step forward into unknown experiences.

Chant 13 MO-RAA-AH

These are powerful feminine vibrations to balance male and female expression. It is used with the 20th ray of lilac and lavender. This is effective in dealing with problems of sexual acceptance and worthiness.

Chant 14 NO-OH-RAH

When using this chant, place an amethyst crystal on the heart chakra. The resonance will activate the soul facet energy of universal law and bring aspects of law into consciousness.

Chant 15 SO-MAA-AH

This is a chant of knowledge, achievement and recognition. It is a mantra to contact the "God within", it moves you to accept your trinity in relation to your God-self.

Chant 16 O-OH-DA

This activates the energy expression relating to your physical involvement in the structure of society, as well as to your physical body.

Chant 17 FAA-RO

This chant is rarely used by itself. Its prime function is in combination with other chants. It adds a power resonant completion to the chants.

Chant 18 ZAU-RAAM

The resonance of this chant activates the soul memory connecting to past incarnations.

Chant 19: KO-FAA-ROO

The chant responds to planetary Goddess energies that change the frequencies of the second and base chakra.

Chant 20: RU-AH-SHIM

This is a sound for magnetic energizing and of your body when the energies are weakened or depleted. By using it with the 25th ray of lemon yellow, it opens a channel of communication with the Ashtar Commands.

Chant 21: SHE-MA

Chant with the 34th ray of pale violet. This aligns you with the female aspect of God, Sananda. They are vibrations of love, softness and unity.

Chant 22: TRI-AH

This vibration brings power energy into accepting your own mastery. It is also to be used for conditions not in order in the shoulders and arms.

Chant 23: THU-MAAR

This chant assists with self-growth and development. It is a chant of recognition and echoes throughout the Universe. Use it with the 28th ray. Three stripes of light blue alternating with two stripes of gold.

Chant 24: AU-MA-LAA

A chant of completion and to attain the power to begin the next level of evolution. It is a sound for the protection and shield of your energies from negativity.

Chant 25: A-LEO-U, BA-EO

These are used to prepare the vibrations for meditation and learning. They serve to unlock energy and allow for the spiritual flow to commence. They are used with the 27th ray of silver pink.

Chant 26: DOU-LAE, VO-A-A

This relates to the energies and the area of the respiratory system. Use it with the 16th ray of yellow-rose light.

Chant 27: ZOO-UR, CHO-RA

Activate the 19th ray of pale green and gold to help you recall past life experiences and incarnations.

Chant 28: VO-A-A, ZOO-UR

This chant assists you in releasing blocked emotional expressions to relieve unwanted pressure that causes distress.

Chant 29: LAU-RR- U, BA- O

This associates with the diseases pertaining to the lower extremities resulting from circulation problems and muscular disorder. Use the 21st ray of orange-yellow with the sounds.

Chant 30: YO-OOH-DA, HA-JO-HA, VO-A-A, HA-JO-HA

This brings forth the total vibrations of God, the total frequencies of the creator of this Universe. Use it with the 36th ray, the perception of all color. When you are ready to acknowledge the total existence of God and are ready to reach beyond into creation, use this to assist you in your expansion.

Chant 31: NO-OH-RAH, FAA-RO

These are universal vibrations of self-acceptance relating to ego and soul. It is to be used with deep purple for power and incentive to achieve goals.

Chant 32: O-OH-DA, FAA-RO

This chant aligns you with the frequencies of the nature spirits. Use it with the 33rd ray of pale rose to create harmony and spirit communication.

Chant 33: KO-FAA-RO, CHO-RA

Use this chant with children and their life circumstances. It will assist them to maintain a balance within their vibrations. It is especially soothing to autistic and retarded children. Use it with the 18th ray of powder blue.

Chant 34: CO-LAE-AH, FAA-RO

This chant connects soul energies to vibrations of earth. It correlates your energies to adapt to society and emotional life.

Chant 35: CHO-RA, KO-FAA-RO

This resonance balances the love, emotional and magnetic energy meridians in your physical body and centers your chakra system.

Chant 36: CHE-MA, AU-MA-LAA

This brings into balance the energies that relate to your female energy expression, and allows you to experience and express softness and compassion to others.

Frank Alper: The 36 Atlantean Chants

All 36 Chants are available with 2 CD set.
www.adamis.ch

Gratitude and Dedication

It is a great joy to finally make "Our Existence is Mind" available to all English readers.

It is also an immense honour to continue publizising and dispersing the "Treasures of Adamis" as Frank called his collection of spiritual wisdom.

I thank all of you who encountered Frank in his life. You surely enriched his life with wisdom, joy and laughter!

With Love and Blessings,

Katharina Alper

Frank Alper: Exploring Atlantis

This is a new and complete edition of the trilogy "Exploring Atlantis" presenting Frank Alper's spirituality through his soul channelled lectures during the time period of 1980 through 1986 in Phoenix, Arizona.
A unique transmission of knowledge about the ancient Atlanteans' social, moral, sexual and spiritual customs and habits by Universal Masters such as Strength from Atlantis, Atemose II, Kryon, Pythagoras and others. A large part of the book is dedicated to the function and use of quartz crystals by the Atlanteans. Patterns of crystals for healing applications are explained in all three sections of this book.

ISBN 978-14475-4994-9 printed by Lulu
www.adamis.ch

Soul Plan

Frank Alper's: Spiritual Numerology written by Blue Marsden

Soul Plan is a new interpretation of an ancient system of life purpose analysis. This method works on the conscious and unconscious level and promises to bring the recipient greater freedom, connection, satisfaction, healing and life purpose activation.

Soul Plan will introduce what, for most people at this time, will be an entirely new and fascinating way of seeing their lives. This system uses the sound vibration and intention behind naming, to determine a person's entire 'Soul Plan.'

Simply by discovering your Soul Plan as revealed by a practitioner, you will naturally begin to align with it. The more subtle message is that in so doing, we may come to realize our true nature. More and more people are now coming to this realization and Soul Plan can be a significant catalyst in deepening this process. When this fully occurs there is a sense and experience of our Soul Plan unfolding naturally.

- Do you know your life purpose?
- Are you still searching?
- Do you have a sense there is something more for you in this life?

Blue Marsden is founder and director of the Holistic Healing College, the London School of Qigong and the London College of Hypnotherapy.
Since a teenager he studied Eastern Philosophy and Jewish Mysticism. After many years of working with various intuitive methods he came across a little-known system that resonated more than any other and after incorporating this into his own practice decided to introduce it as a module in the counseling program he founded. Over the years Blue researched the origins of this work and on discovering synchronicities with his own non-dual understanding has subsequently modernized it, added

new interpretations and channeled additional material and healing interventions to form what is now known as the Soul Plan system.

During the early and mid-1990s he explored the use of hypnosis and sound healing to aid and enhance performance and creativity. He specialized during this time in working with artists, singers, writers and presenters.

In his psychotherapy practice he specialized in relationship counseling and was asked by the British Association of Anger Management to produce an audio program later acquired by many schools across the country.

He is also a teacher and long-time practitioner of chi kung. This discipline helped him to gain a deeper perspective on the rarely recognized 'magnetic' and health-enhancing quality potentially present within the 'talking cures'.

Blue holds a degree in Philosophy and a Postgraduate Masters degree in Psychoanalysis and his work has featured in the national media.

ISBN 978-1-78180-076-8 published by Hay House

www.soulpan.co.uk

For Personal Notes

www.ingramcontent.com/pod-product-compliance
Lightning Source LLC
Chambersburg PA
CBHW070657100426
42735CB00039B/2222